The Treeline

The Treeline

*The Last Forest and the Future
of Life on Earth*

BEN RAWLENCE

JONATHAN CAPE
LONDON

2 3 5 7 9 10 8 6 4

Jonathan Cape, an imprint of Vintage, is part of the Penguin Random House group of companies whose addresses can be found at global.penguinrandomhouse.com

Penguin
Random House
UK

Copyright © Ben Rawlence 2022

Ben Rawlence has asserted his right to be identified as the author of this Work in accordance with the Copyright, Designs and Patents Act 1988

First published by Jonathan Cape in 2022

penguin.co.uk/vintage

A CIP catalogue record for this book is available from the British Library

ISBN (hardback) 9781787332249
ISBN (trade paperback) 9781787332256

Illustrations by Lizzie Harper
Map by dkb creative

Typeset in 11/15 pt Sabon LT Std by Jouve UK, Milton Keynes
Printed and bound in Great Britain by Clays Ltd, Elcograf S.p.A.

The authorised representative in the EEA is Penguin Random House Ireland, Morrison Chambers, 32 Nassau Street, Dublin D02 YH68

Penguin Random House is committed to a sustainable future for our business, our readers and our planet. This book is made from Forest Stewardship Council® certified paper.

MIX
Paper from
responsible sources
FSC® C018179
www.fsc.org

To the trees and all living things that call the forest home

Contents

THE BOREAL FOREST & THE ARCTIC TREELINE

Prologue

Yew, *Taxus baccata*

Llanelieu, Wales
52° 00' 01" N

Behind my house is a very large and very old tree. I never gave it much thought, a commonplace thing, a gnarled old tree by a churchyard, a typical Welsh scene. But lately I have found myself paying more attention to trees.

The tree in question is a yew, *Taxus baccata*. It stands on a mound several metres above the road, roots tightly gathered below the soil, bunched muscles under skin. The yew's delicate evergreen needles resemble fine hair and they hang from great curved branches in an untidy fringe hiding a face – a shy Green Man perhaps. To approach the trunk, you must duck your head beneath the swooping fringe and part the branches like heavy sacred curtains, as if venturing behind an altar. It is a mysterious refuge from the path only steps away, rich with the acid tang of evergreen, of life.

On the opposite bank of the path is another yew, slightly smaller but with the same smooth pinkish bark, furry and sticky in places. I follow its exposed roots bursting from the soil, snarling their way along the bank and under the path, entangling with those of its larger neighbour, forming one living structure. Upon closer inspection, the smaller tree is sporting bright red berries: she is female. The larger one, without fruits, is male. They are a handsome, imposing pair, but try as I might I can't find anyone who knows how old these ancient lovers are, nor how they got here.

Dating yew trees is notoriously hard. This is partly because there is no upper age limit. They grow rapidly in youth, steadily in middle age and can survive in senescence for an apparently unlimited period. Sometimes growth can stop and the tree can stand dormant for long periods, possibly centuries. Tree ring analysis fails with yews. Like cedars, they can grow from a low-hanging branch that has rooted in the soil, and shoots can sprout from stumps; left alone, a yew might be capable of regenerating itself for ever. This was one of the things that made them sacred to the Celts. They worshipped the yew with its toxic red berries, pink flesh and copious sap precisely for its godlike attributes, its ability to bestow life and death and for its claim to immortality. The churchyard is circular, an indicator of a *llan* – a pre-Christian sacred site preceding the little Norman church. Yews are often found with *llan*. The old couple standing quietly above the stone circle, holding hands under the path for centuries if not millennia, might be the reason the village of Llanelieu is here at all.

Ancient trees are a source of wonder. Refugees from another era with a life cycle so much longer than human timescales. Their distribution and range is the result of incredibly long planetary cycles of geology, climate and evolution. The curious distribution of yews, for example, found only in the high mountains of Central Asia and scattered redoubts of northern Europe, suggests that it must once have been more widespread and is now a relict species – the remaining examples are outliers from a different epoch. This may be a consolation in moments of crisis, a reminder that our concerns are mere specks in the deep accumulated time of thousands upon thousands of tree rings. But now that humanity has upset the planetary systems of oceans, forests, winds and currents, the balance of gases in water and air that gave rise to our species, their consolations are in question. Trees no longer offer comfort, but warning.

It is our complacent attitude to time that is the first casualty of global warming: millennia have become moments. These days I cannot look at the mountain, the forest or the field without feeling the ground tremble in both anticipation and memory. Our best guide to the coming uncertainty is history: geology, glaciology and dendrochronology – the studies of rocks, ice and trees. Thus, the past and the future are made immanent, time has become slippery, and a walk in the hills can make you dizzy. Suddenly I see trees everywhere: where they are not, where they have been, where they should be. It is a way of looking at the landscape outside of time, as people closer to the earth have always done. And, seen as such, the view looks wrong. The clean, green lines of the Black Mountains that rise above the church and the village now appear to me a tragic desert, a monument to a geological epoch of collective human folly.

These hills are the border between England and Wales. The crossing of this line first by the Romans, then later by the Danes and then the medieval kings of England marked the beginning of a movement which is finally reaching its endgame in the last great vestiges of wildwood on the planet: the tropical Amazon and the subarctic boreal. The Romans, Danes and the nobles of England were in search of natural resources, principally timber. The colonisation of Wales was the first expression of an economic system founded on overreach: having exceeded the limits of what their own environment could sustain, early mercantilists applied force to acquire tribute and resources elsewhere. Empire, whether British, Viking, Roman or otherwise, is by definition overreach. And colonialism, capitalism and white supremacy share a common, perverse philosophy: limits on some humans' freedom of action are seen as an affront to the principle of freedom itself. The exact opposite of the co-evolutionary dynamic of the forest.

Once upon a time, these hills were covered in trees. All that's left now is a patchy ecosystem called *ffridd* or *coedcae* – hawthorn, scrub and bracken mixed with broadleaves – a transition zone between lowland and upland habitats. The peat on the top is testament to the forest that once was. But that was before our neolithic ancestors cleared the forest for grazing and fuel, and before our later penchant for deer, grouse and of course sheep. Before the trees, however, before there was any covering on the rock at all, there was ice.

The last ice age ended 10,000 years ago, mere seconds on the planetary clock. The old yews of Llanelieu could be the grandchildren or even the children of one of the first trees that took root as the ice retreated. Conifers like yews have evolved specifically in relation to the cycles of ice. They thrive in marginal environments, in tough soil with limited nutrition. This is the process of the treeline at work. For the treeline is not really a line at all.

The fact that in modern usage the term 'treeline' has come to mean a fixed line on a map indicating the growing limit of trees is simply evidence of the very narrow time horizons of humans, and of how much we have come to take our current habitat for granted. In fact, the growing conditions for trees, whether limited by altitude (up a mountainside) or latitude (towards the North Pole), are only as certain as the environment that produces them: the availability of soil, nutrients, light, carbon dioxide and warmth. For a couple of millennia these climatic conditions have remained remarkably constant, but over longer timescales tiny changes in global temperature have meant that the treeline has always been a moving target.

The ice has come and gone many times. And each time nature has begun again, slowly recolonising the land scoured of soil. First comes lichen, then moss, then grasses, shrubs and the pioneer trees like birch and hazel that improve the soil and dump tons of leaf litter for the slower-moving greats that follow: the pine, sessile oak and yew. Left to its own devices, nature's equilibrium in most habitats on earth unless limited by cold or drought tends towards the eventual production of forest. And thus, as the ice moved north, the treeline slowly followed, taking root in meagre soil, photosynthesising, shedding its needles, then dying to create the rich fertile crust of the earth, laying the foundations for the habitats of all other terrestrial life. There is scarcely a patch of the northern hemisphere over which the treeline has not passed.

Ever since the Pliocene epoch, three million years ago, when the explosion of plants cooled the atmosphere to its modern equilibrium, ice ages have marked our planet in 100,000-year pulses. The pulse is because the earth does not spin evenly but wobbles like a top. The wobble is called the Milankovitch cycle. It tilts the planet a fraction away from the sun every 100,000 years, chilling it ever so slightly and causing the ice at the poles to expand and retreat in a millennial version of our annual seasons. The South Pole is an island and glaciers are rare in the southern hemisphere apart from New Zealand and Patagonia. The northern hemisphere meanwhile has been forested and deforested over and over. Time-lapse photography of geological time on planet earth would show a sheet of ice descending and retreating in a rhythmic pattern, and a green mass of forest rising towards the North Pole then falling again, like breath.

But now the planet is hyperventilating. This bright green halo is moving unnaturally fast, crowning the planet with a laurel of needles and leaves, turning the white Arctic green. The migration of the treeline north is no longer a matter of centimetres per century; instead it is hundreds of metres every year. The trees are on the move. They shouldn't be. And this sinister fact has enormous consequences for all life on earth.

I can't remember where or when I first heard about trees on the march. But the image stayed with me for several years before I took the trouble to research what was actually going on. I had assumed scientists had observed minor increments of change, perhaps in response to recent warming trajectories over the last few decades. I was totally unprepared for what I discovered.

I learned that the Arctic tundra is getting shrubbier, turning green; but this is not a simple story of trees gorging themselves on carbon dioxide and racing north. It is a picture of a planet in flux, of ecosystems adjusting to massive changes and trying to find their balance. Of forests the size of nations being destroyed by fire, parasites and humans every year while elsewhere precious tundra is colonised by trees now rendered as invasive species. Forests are evolving their communities of species or popping up where there should be none, creating havoc for those animals and humans whose survival strategies relied on them staying put.

Our maps are out of date. The position of the Arctic treeline has been one of the definitions of the Arctic Circle. It almost exactly tracks another, the ten-degree July isotherm – the line around the top of the world marking an average summer temperature of ten degrees Celsius. This wavy line briefly touches the tops of the Cairngorm Mountains in Scotland before making land again in the interior of Scandinavia away from the temperate forested fjords. From the plateau of Finnmark, it then runs in an unbroken line from Russia's White Sea across the top of Siberia to the Bering Strait. In Alaska the treeline pushes up against the Brooks Range before taking a diagonal plunge across Canada, meeting the sea once more at Hudson Bay. On the other side of that inland sea it wends its way through Quebec and mountainous Labrador and then makes the leap to southern Greenland.

This is the route of the journey described in this book, but the concept of a line itself is misleading. Zoom in, and the treeline is not a line at all but a transition zone between ecosystems, what scientists call the forest–tundra ecotone (FTE), in some cases hundreds of kilometres wide and in others a matter of feet. As the climate warms, the zone and the huge ecosystems of tundra and forest on either side are being transformed in diverse and unexpected ways. And anyway, the line is wrong. The ten-degree July isotherm is no longer a stable fact upon which cartographers can rely; it swings wildly all over the place, as summer temperatures in Siberia, Greenland, Alaska and Canada can attest. Where trees are able to grow and where they actually are now has become steadily uncoupled. This makes the whole area a zone of possibility, and of threat.

Journeying along the zone, I learned much about the fundamental role the northern forest plays in regulating earth's present climate. More than the Amazon rainforest, the boreal is truly the lung of the world. Covering one fifth of the globe, and containing one third of all the trees on earth, the boreal is the second largest biome, or living system, after the ocean. Planetary systems – cycles of water and oxygen, atmospheric circulation, the albedo effect, ocean currents and polar winds – are shaped and directed by the position of the treeline and the functioning of the forest.

I learned how little we know about the changing operation of these systems under warming. We know the world is getting dangerously hotter; we don't yet know what that will mean for us or the other life forms of the forest. As they warm, forests are losing their ability to absorb and store carbon dioxide. While the boreal is the greatest planetary source of oxygen, more trees there does not necessarily mean more carbon sequestered from the atmosphere. As trees invade frozen tundra they hasten the melting of permafrost, frozen soils that contain enough greenhouse gases to accelerate global warming beyond anything scientists have modelled. Many contradictory things are happening at the same time.

The earth is out of balance, and the treeline zone is a territory in the grip of large geological change, confounding and challenging our ideas of the past, present and future. 'We are in between stories. The old story, the account of how the world came to be and how we fit into it, is no longer effective. Yet we have not learned the new stories,' says the cultural historian Thomas Berry.[1] I found the seeds of those new stories rooted in older arrangements in the boreal. For the most part, forests are places where human ways of coexisting with nature on equal terms still persist.

The terrain, both scientific and geographic, though, is vast, and the scope of what the boreal represents so huge, it seemed impossible to encompass within the scope of a single book. It was only when I discovered that a tiny handful of tree species make up the treeline that I began to see that an attempt at description might be possible. An elite club, the six featured here are the familiar markers of the northern territories: three conifers and three broadleaves evolved to survive the cold. Moreover, remarkably, each of these tree species has made a section of the treeline its own, outcompeting other species and anchoring unique ecosystems: Scots pine in Scotland, birch in Scandinavia, larch in Siberia, spruce in Alaska and, to a lesser extent, poplar in Canada and rowan in Greenland. I decided to visit each tree in its native territory, to see how the different species were faring in response to warming, and what their stories might mean for the other inhabitants of the forest, including us. My visits to the different places occurred between 2018 and 2020 at different times of year to capture the seasonal workings of the forest, but the chapters that follow are arranged geographically, tracking the treeline east, towards the rising sun.

These northern species are few, but they are tough. In the long game of geological natural selection, only the most creative survive at these latitudes of extreme cold. While the delicate, biodiverse, tropical rainforests may have maintained a familiar assemblage of species for millions of years, the more northerly latitudes are slates that have been wiped clean again and again. This is the place to look for a glimpse of what, after the great transformation currently unfolding on earth, will remain. Thousands or perhaps millions of years from now, when the planet cools again, the species that creep out to repopulate the earth could well be those that are endemic to the boreal. They are uniquely adapted to climate change. They have been riding the tides of ice for millennia. Deforestation and existing emissions in the atmosphere have already doomed much of the world's rainforest to savannah. My neighbours, the old Green Man and Woman of Llanelieu, might make it, depending on how hot and dry the island of Great Britain becomes, and depending on the scale and success of human efforts to limit the damage, but the last forest will be the boreal. When humans are only fossils, it is these hardy northern species that will still be standing tall.

1. The Zombie Forest

Scots pine, *Pinus sylvestris*

Glen Loyne, Scotland
57° 04' 60" N

As the ice retreated to higher ground at the beginning of the current interglacial period, the boreal forest set off in pursuit. Plants that had not been seen on the islands of Britain for thousands of years began, gradually, to return. Ice persisted on the uplands of north Wales and the Highlands of Scotland, but in the valleys and the plains, lichens formed a crust on the exposed rocks. Then came mosses with their creeping fur, laying the ground for grasses and sedges first, soon to be followed by the pioneer shrubs of hazel, birch, willow, juniper and aspen. This boreal system worked its way north, across the land bridge where the English Channel now is, a sweeping tide of green on the heels of the ice, the cocktail of early seeds dispersed according to the natural cycles of wind, rain and the migratory patterns of animals, including humans.

Ten thousand years later, I follow. Pointing the car north from Wales, I head to where the map says the treeline has come to a halt at its present position: Scotland. Driving to Fort William through the spectacular soaring valleys along the west coast of Scotland, the rocky outcrops of the peaks appear stationary, like the roof of a cathedral merging with the sky. The rich green slopes roll back and forth with every bend in the road; scree tumbles in long runnels like waterfalls from hidden lakes of rocks high above. Sunlight shears the view, one minute blinding, the next revealing a promised land.

It is not until I am actually there that the contradiction strikes: I am searching for the upper limit of the forest, but where is the forest? Scotland's forbidding hills, rank upon rank of shadowed slopes rising out of the mist, are such a durable sight in collective memory and culture it is almost impossible to imagine them otherwise, and yet Britain was once, briefly, an island of trees. Caledonia, as it was named by the Romans, means 'wooded heights', but its 'great wood' has become a mythical thing. Scotland's bare hills are both epitaph and warning: this is where the commodification of nature leads.

To ask what is happening to the treeline in such a ruined landscape is a profoundly political question. On paper, Scotland is held to be the southern and western limit of the Arctic treeline in Europe; estimates based on temperature and growing seasons suggest that here it should be at 700–750 metres.[1] Stumps have been excavated at 790 metres dating from a slightly warmer era 4000 years ago.[2] But how the treeline is responding to warming now is hard to say because nearly all the trees were cut down. Efforts to restore Scotland's great wood are under way, 're-wilding' the hills and planting trees, partly to allow them to find their level and re-establish a natural transition zone between the forest and the moor. But such changes are controversial. How we see the present and the future often depends on our understanding of the past. What is natural? What is being restored? Meanwhile, as humans debate ecological history, global warming gathers force, threatening to render our meagre response irrelevant.

The treeline's first wave, or primary, vegetative cover after the last ice age resulted in a patchy forest that the foremost historian of British landscape, Oliver Rackham, calls wildwood.[3] This was a dynamic shifting community of plants – at its southern end connected to mainland Europe by the land bridge, and at its northern frontier petering out into the moorland tundra of the 'flow' country in the far north of Scotland and the scattered rocks of the Hebrides, where the dry cold of the Arctic polar vortex wrestles with the Gulf Stream for influence.

This wildwood was rampant but precarious. Birch was quick to establish but transitory, giving way to other, bigger and bolder trees. As the evolving society of the forest worked out its own logic, a steady state would emerge with a particular tree or trees dominant. In much of southern England this was lime, in the north and Wales it was a mix of hazel and oak. In the Highlands of Scotland the apex tree was originally oak. But the steady state of the wildwood could be upset and tipped into another cycle by an influx of a new species or a change in the weather. The introduction of the pine was one of these.

Around 8500 BCE pollen records show Scots pine (*Pinus sylvestris*) arriving suddenly across Britain, colonising a corridor up the west coast of the British Isles, nosing its way into the inlets and fjords of Scotland and then across the straths and valleys and up into the mountains. Pine out-competed the birch and oak that had generously worked up sufficient soil for it to flourish. So successful was the pine that the birch disappeared almost completely for thousands of years, surviving only in a remnant zone in the flow country north of what is now the city of Inverness.

This pine wood spread across Scotland, reaching its apex of round 80 per cent of the land area, according to Rackham, around 4500 BCE. Recent archaeology, pollen analysis and even the 7000-year-old bones of pine trees preserved in bogs have fed debate about the scale and the fate of Scotland's once magnificent wildwood.[4] Conservationists are seeking a record to guide their attempts at 'ecological restoration'. Opponents are seeking evidence that the trees were eliminated through natural causes and that the current status quo of grouse moors and deer parks is just as deserving of the designation 'natural'. At issue, it seems, is one vision of nature over another, neither of which attributes much influence to humans for creating the shape of the landscape in the first place, and yet the history of humans and the history of the forest is deeply entwined.

Before driving north, I read a scientific paper by Lithuanian researchers demonstrating that the DNA of the Scots pine in the eastern half of Scotland came from a refugium – a place where species survived the last ice age – near Moscow around 9000–8000 BCE.[5] Previous DNA analysis has shown that the surviving pines in the west of Scotland came from the Iberian peninsula in modern-day Portugal and Spain. In both cases the seed migrated to Scotland on timescales hundreds of times faster than is possible through natural succession. The most likely vehicle for such rapid migration was humans.

There is a myth in Celtic folklore – with an apparent grain of truth – that when the Celts colonised Scotland they met Ukrainians coming the other way. For the Celts, the pine was a sacred tree with a myriad of uses. The pine was *ailm* in the Celtic alphabet, the ogham script, and it is very likely that they brought it with them from Ireland and Wales. It was perhaps sacred too for the mysterious Ukrainians, who were part of the Celtic kingdom, 'the people of the Danube' in old Irish, the only others with red hair. For humans so tied to nature and reliant on plants it would make sense to travel with your own habitat. Something twenty-first-century humans might soon wish we were able to do.

The result, in the present day, is two distinct genetic communities of Scots pine in Scotland divided by the Highlands. They have yet to cross-pollinate and conservationists are keen that they do not since the genetic and chemical distinctiveness has consequences for other species that rely on the keystone of the pine. Insects like wood ants, for example, can taste differences in resin and will choose particular trees. Leaf chemistry, flower timings and growth forms are all different. The crested tit remains east of the Cairngorms, embedded in its environment. However, the conservationists needn't worry yet. The risk of interbreeding is minimal since the fragments of surviving forest are spread out and very small. Less than 1 per cent of Scotland's old-growth pine woods remain.

Rackham argues that the pine wood never stretched from shore to shore, but it certainly covered most of Scotland until Mesolithic humans began to clear the forest for agriculture, hunting and construction. Managing the forest through felling, clearing or burning for game played a role in creating biodiverse habitats of heath and moor, but also set the stage for the creeping blanket bog that has become upland Britain's signature landscape. The bog is, in a sense, a ruined ecosystem as tree clearance has allowed minerals and iron to be washed into the lower layers of the soil, creating a pan impermeable to water. Unable to drain, the tundra-type landscape becomes waterlogged, and plants do not fully decompose, forming peat.

The pastoralist indigenous crofters, who farmed the Highlands till the clearances of the eighteenth and nineteenth centuries, traditionally moved their cattle between the lowland forest and the moor. The clearances and the subsequent expansion of Victorian shooting estates for grouse and deer are often blamed for the deforestation of the Highlands, but while heather burning and overgrazing by deer in the absence of apex predators like wolves, lynx and bears did indeed prevent the trees from coming back, much of the open upland landscape had already been formed by clearing all the trees.

Traditional custom and practice, inherited from the Celts, respected the woods. Pine was a renewable source of building materials, fir candles for light, tar and resin for tanning and waterproofing, fibres for ropes and bark for kindling, flour and medicine. Until well into the 1960s, pine sap provided tallow for candles, forest timber was used for railway sleepers and boats, and pipes were made from hollowed-out trunks. Indigenous systems apportioned a host of rights to goods provided by the forest – hazel rods, firewood, timber, mushrooms and animal fodder – and there were strict moral and financial penalties for wasteful coppicing, for unsanctioned pannage (the grazing of animals in common woods) and so on. As many other more recent episodes of tropical deforestation show, indigenous use of the forest is often the most reliable form of conservation. The so-called tragedy of the commons (that humans cannot be trusted to manage a common resource sensibly) might be a problem for individualistic societies unable to restrain pollution and over-exploitation, but as a historical explanation for the British landscape it doesn't hold except perhaps as a retrospective ideological justification for the real tragedy to follow: the enclosure of common land.[6]

Property rights over land was originally a Roman idea resisted by both Greeks and Celts, who maintained that nature could never be owned by humans, only used. Hundreds of years after the

Romans left Britain this notion cleared the path for foreign land-
lords and the extreme concentration of land ownership in Scotland
today.[7] The woods had been used by the clans. They needed the
forest. Indeed the word 'forest', and its endurance on maps despite
the lack of any trees, is an echo of its earlier meaning as an unfenced
area protected for hunting and common use, more latterly by the
Crown. The shift from rights of usage to rights of ownership, seen
as the mercantile spirit of northern Europe inveigled or imposed
itself across the world, was, it seems, the crucial shift, as forests
ceased to be seen as sacred places of wonder, mystery and susten-
ance and instead became a standing crop with a value expressed in
pounds, shillings and pence calculated by the acre and the ton.

Scotland and Ireland and their natural resources, foremost
among them their remaining timber, were the front line of that
early capitalist desire that expressed itself in colonialism. English
kings needing ships, houses, carts and cathedrals from the medieval period onwards – well before Henry Hudson and John Davis
were dreaming of the North West passage and Sir Walter Raleigh
of the Orinoco – first looked to Wales and then their colony of
Ireland. Then, with the Scottish and English crowns united and
Ireland's woods gone, it was to Scotland that they turned.

Along Loch Linnhe, low cotton-candy clouds scud between the
peaks of Ardgour across the water. The peninsula holds the most
south-westerly remaining relict pine wood, within the estate of
Conaglen, a property given over to deer stalking. In a hollow
between hills lies the last scrap of forest to which former generations owed so much of their wealth. An observer commenting on
the huge amounts of timber being imported into Ireland from
Scotland wrote in 1686, 'there used manie shipps to come to that
countrie of Ardgoure, and to be loaded with firr jests, masts and
cuts. This glen is verie profitable to the Lord.'[8]

The green hills sheer into the black depths of the loch. A train
rattles along beside the water heading for the end of the line. The
wealth of the forest even shaped the geography of Scotland. The
River Spey was dammed and altered to facilitate the floating of

logs to the sawmills and shipbuilders of Speyside until steam rail-
ways put the floaters with their particular vocabulary and their
currachs – a light frame boat covered with hide for the return journey upstream – out of business. On the west coast the timber came
out along General Wade's military road and then the railway that
terminates at Fort William at the head of Loch Linnhe.

Beyond Fort William, the famous Road to the Isles opens up.
Majestic sonorous glens beneath blue peaks fall away to face Skye
across the sea at Knoydart. Instead of timeless, this landscape now
appears apocalyptic to me: the victim of a catastrophe. It is perhaps a miracle that any old-growth forest has survived at all.
However, due to the odd enlightened laird, far-sighted forestry
official or sheer remoteness, eighty-four fragments of native Caledonian pine wood remain. These are the 'granny pines', gnarled
and apparently half-dead characters that animate the otherwise
blank canvas of some Scottish hillsides. The oldest known speci-
men is 540 years old and grows in a remote boggy valley called
Glen Loyne. These are the only trees that were large enough to
escape browsing after the wolves were extirpated and the deer and
sheep allowed to run riot.

There is something profoundly wrong with a solitary pine. Pines
are social creatures; they rely on other trees for sharing resources
through fungal networks. When mature, pines transport carbon
underground to support young saplings, and in old age carbon and
nutrients travel in reverse, the young trees helping out the older
ones. The natural lifespan of a Scots pine is up to 600 or 700 years
within the healthy network of a forest. Scotland's surviving granny
pines are mostly under 400. Major dips in the pollen record suggest this is because of the massive extraction of trees from 1690 to
1812. According to dendrochronologist Rob Wilson, 'You can still
see the effects of the Napoleonic Wars in the structure of the forest.'
But there is another factor.

Lone trees are prone to sudden dying before the end of their normal life expectancy. Could it be that these matriarchs of our oldest
forests, these stewards of our ancient ecosystems and midwives of
so much industrial wealth are, in their old age, lonely? Native

American stories tell of solitary trees 'speaking' to humans of their loneliness, asking people to plant them neighbours. Are the granny pines missing the companionship, and the meals on wheels, provided by their children? Are they mourning the ghost of the forest?

I leave the car at a lay-by on the single-track dead-end road. The valley unfolding below holds the story of the latest iteration of Scotland's encounter with industrial capitalism, a widescreen landscape of destruction. The hillside opposite has the camouflaged look of 'muir-burn' – uneven brown and fawn stripes created by heather burned to encourage grouse for shooting. This would otherwise regenerate into woodland and looks like a roughly shaven scalp. Lower down are the scars of a mono-crop spruce plantation, a comparative desert devoid of biodiversity; the dark military-green trees planted too close together to allow any other life. The hillside has been ravaged where machines have felled trees before they have reached maturity, leaving great brown furrows down which the precious topsoil flows in boiling ribbons to the loch. Further along is a derelict larch plantation that someone forgot to thin, the trunks branchless, half the trees fallen and collapsed in on each other, wind-thrown because they have no root strength. On the loch bob pontoons, and between them the surface of the water is stitched with the floats of an intensive salmon farm. Plastic barrels of feed are piled six metres high on the bank. And above them steel pylons fizzing with sixty-six kilovolts march along the shore all the way to the hydroelectric power station at Kingie, a red and yellow hard-boiled sweet atop a concrete dam. Even the loch is artificial: the landscape as the ultimate resource, seen only through the bloodless eyes of an accountant. There is nothing natural in the view apart from a clutch of willows by the stream in the foreground.

A footpath sign points north, uphill. Affixed below is a warning:

TAKE CARE: You are entering remote, sparsely populated, potentially dangerous mountain country. Please ensure that you are adequately experienced and equipped to complete your journey without assistance.

On the other side of the glen begins one of the remotest stretches of wildernesses in Britain; you can walk for three days to Knoydart without sight of a road or house. It is why the granny pines of Glen Loyne are still there. Getting them out of the glen was such hard work, they were left till last, then probably forgotten.

The muddy track wends its way between a stream and a deer fence. There is one other set of boot prints in the mud. They are not fresh. Inside the fence, birch, willow and pine seedlings seem to be doing well. It looks odd, a boxed area three times more overgrown than the rest of the hill, but outside the fence are sheep and deer. This is the front line of ecological restoration in Scotland, the divide between those invested in an economy and a landscape based on commercial forestry and shooting, and conservationists committed to defending the trees from being eaten. The struggle has all the passion – and barbed wire – of a war.

Soon I am clambering up steps and over stones. Little butterwort with their Venus flytraps cling to the rocks, and a huge club moss like a beaver-skin hat, fifteen centimetres thick, out of which grow alpines and grasses and other mosses with fine wispy hairs like an old man's beard, sits on top of a boulder.

The boulder is a miniature showcase of the treeline at work: a future hummock. An exposed rock will first be colonised by crustose lichen mining its minerals, growing at a rate of 0.1 millimetres a year. To obtain the minerals, the lichens secrete an acid which breaks down the rock. Other foliose lichens, with more leafy structures, take advantage of this broken layer as do mosses which then trap further organic material in their fronds and accelerate the process of soil accumulation. When this top layer eventually overcomes a tree stump or boulder and joins up with the surface soil, it will form a hummock. Hummocks in regular sizes close together are often remnants of ancient woodland where soil has accumulated over tree stumps. Their formation can take decades, if not hundreds of years. How many seasons has this soggy household seen?

By the time I reach the ridge I am thirsty as hell. I had expected to find burns all over the hill and so felt no need to carry water, but

I hadn't reckoned on the blanket bog. Britain has 13 per cent of the world's peat, much of it degraded and fast drying out. The peat accumulates like a creeping wet lava field, a few millimetres every year. Once the trees are gone from a landscape like this, it is hard for them to return, and so, from the top of the ridge, I can enjoy an uninterrupted view of clipped green glen from the head of Loch Loyne in every direction. Looking north-west, mountain after mountain rises in an undulating vista of granite and grass. There is no sound except the wind tugging at my hair, no birds, no trickle of water. It is easy to see why the British ecologist Sir Frank Fraser Darling famously called the Scottish uplands a 'wet desert'.

In search of water, I thrust my arm into a tufted gash in the peat, deep under the skin of the hill almost up to my shoulder in order to raise a meagre cup of brown tannic liquid. It tastes bitter but it'll do. Oh, for the multitudes of forest filaments to filter the water table into a clear sweet drink!

Descending out of the wind into Glen Loyne, the sound of rushing water rises to greet me. This is the faint roar of the rapids that I can see striating the river far below as it carves its lazy thousand-year sweep into the bowl of the valley. The sound is a reminder of how alone I am. The view is a surprise, like coming upon a hidden African savannah behind a familiar hill. A lone rowan rises from a crack in a boulder, out of the reach of deer. Further down, there is the faint line of a broken deer fence, and then the whole glen opens up. A carpet of green up to jagged ridges at 600 metres. And in the foreground, supposedly protected by the fence, hundreds of widely spaced ancient pines stretch away into the distance.

It is as if I am the first person to stumble upon the aftermath of a battle. There is a preponderance of grey-white trunks called snags, like standing skeletons. Other ancient trees are half green with needled limbs scrabbling at the air, stripped of flesh like zombies stumbling up from a tomb.

The oldest in the enclosure is not the tallest. I call her she, even though pines are monoecious with male and female parts to their flowers. There is a blaeberry plant growing in the crook of one of her branches, and ferns sprouting all along the length of another. Red, orange and black spotted lichen cover her flaking pinkish bark, and strands of green horsehair lichen hang down like cobwebs from her extremities. In a flourishing forest these would once have formed dense sails for trapping moisture. Her branches droop and sweep away from the prevailing wind, tapering to short spiky green needles, each finger ending in a thick dusty brown 'candle' about the size of a cigarette – the growing tip of the branch. High up on her trunk are dark holes, some flecked with recent droppings: homes for owls or woodpeckers.

There is a pathos to this giant tree providing a habitat for others to rear their young and reproduce while its own offspring are sacrificed annually. The remnants of seedlings lie shredded on the leeward side of the tree, ravaged by deer that have broken through the fence. Deer are woodland creatures essential to the healthy functioning of a pine wood – opening up ground through grazing, fertilising the forest floor with their dung – but if they stay too long or become too many, they can wreak havoc. Deer will eat almost any tree as high as they can reach, and they will attack seedlings with their antlers to scratch them, breaking young stems in half. The Scottish nature writer Jim Crumley calls the wolf the 'painter of mountains' because it keeps deer populations in check.[9]

Deer droppings are everywhere beneath the spreading canopy, and as I walk through the enclosure, everywhere among the heather too. The chewed-off branches of young birch, rowan and pine lie like stripes across the hillside. For 540 years this granny pine has been setting seed in the hope of offspring. In the early years of the fence, perhaps, seedlings here were able to establish, but I doubt any of them will survive another winter now.

The pine's nearest neighbour is a single bone-white trunk standing like a totem pole. A faded blue tag nailed to its body reads '50a'. A terrible feeling hangs over the land. The pines are like wounded soldiers frozen in the act of falling. The granny pine has felt the brush of a lynx's fur or the wet nose of a wolf, has seen her neighbours felled for ships to fight Napoleon. A mighty tree that can withstand the worst of weather and disease but cannot protect her young against the appetite of deer.

She knows what is happening. Monoterpenes are volatile organic chemicals produced by pines that the trees use to send signals to each other – to deter herbivores or insects or to coordinate seed release. Monoterpenes are tiny molecules that carry pine scent and bounce sunlight back into space. When pines are metabolising in sunlight there can be as many as 1000–2000 particles per cubic centimetre in the air around the tree, reducing the amount of solar radiation hitting the earth. Via the density of the chemical signal and the availability of light, they can detect the presence of other trees. In fact, they see space in polygons, growing away from their neighbours and towards the light, creating a five-sided tessellation in the canopy that is the basis of self-organisation in the forest.[10] Through the structure of their cells, trees can capture reverberations and 'hear' sounds around them as well as ultrasound far away.[11] Pines can detect the familiar presence of rustling needles or the crack of a falling tree, and of course they communicate and look after each other through the rich mycorrhizal network underground. Scots pines have one of the most developed fungal networks in the soil, with over nineteen known ectomycorrhizal relationships for sharing carbon, nitrogen, essential acids and other nutrients.

Indeed, all around me, looming out of the bog, huge fecund mushrooms form rings around dead stumps – the forest genome persists below the soil, waiting. The trees might have to wait a few years yet, but will there be time? With the legions of standing dead dotted around the head of the valley, the number of elderly trees sporting one withered arm, meagre needles and a small crop of stunted cones, it is hard to avoid the feeling that the granny pines are on the verge of giving up. I can imagine them shrugging their gnarled shoulders on their lonely hillside and saying 'What's the point?' Whatever chance Britain's oldest pines have of passing on their genes to the next generation is only as real as the resolve of humans to maintain the fence.

Loch Maree, Scotland
57° 42' 37" N

There are very few places in Scotland deer cannot reach. The
islands in Loch Maree are one of them. From Glen Loyne it is a
long winding drive north through sparsely populated glens past
several other fragments of Caledonian pine forest: Attadale, Tao-
dail, Achnashellach. But even these have a history of logging and
disturbance. The islands of Loch Maree are unique in that they
have been continuously wooded for nearly 8000 years, only
inhabited by the occasional mystic or monk – the ruins of a priory
sit on one of them. Deer can swim of course, but the islands are far
from the shore and while deer have been known on the islands,
none have done meaningful damage.

It is nearly midsummer. The evening is a warm nineteen degrees
Celsius. The sun is still high in a sky blue and crisp as paper. The
loch is a glittering black. The polygonal tower of Bein Eighe looms
above the dark water like an outlier from the Alps or the Himalayas. Its dramatic scree slopes hang at impossible angles. Bein Eighe
is the Eiger of the British Isles, surrounded by a coterie of intimi-
dating Munroes, all impressive summits in their own right. From
the foothills of the mountains, ancient pine woods tumble down
all the way to the water's edge, creating one of the most beautiful
waterfronts in Scotland. Loch Maree stretches like a long slender
finger pointing south-east opening out into a palm at its northern
end before flowing north-west into the sea; a palm holding a clutch
of jewels: its emerald islands.

From the beach at Slattadale, the sun throws a pale fire across
the lake to set the trunks of the pines on the wooded islands alight.
Thick stands of trees are visible in a tantalising pristine paradise, a
nature reserve where camping and motorboats are forbidden. I
stand on the pebbly beach, my toes cooled by the brown rivulets of
peaty waves, contemplating the islands glowing in the golden light.
I came here for the wooded islands, but I hadn't thought about

how to reach them. Beyond a few picnic benches and a Forestry Commission car park there are no facilities to speak of for dozens of miles. No canoe hire shack, no ferry. But I must set foot on the pristine islands. I must inhale the scent of the oldest continuous stand of ancient wildwood, perhaps the only surviving one, in Britain.

The thin line of evergreen canopy atop orange trunks is only just visible a kilometre away. The trees beckon, they dare. There's only one thing for it. It takes me thirty minutes to summon the courage to swim and another thirty to actually make the crossing. It's not the distance that is the problem but my mind. Indeed, it is my mind that begins to let me down halfway across. What if I get cramp? If I tire? In the middle of Loch Maree, no one will know what happened. Three hundred metres of the blackest water lie beneath me. Five hundred metres in either direction. It is 1027 strokes across. I make it. The water on the far side slaps against granite slabs that disappear into the deep. I lie on the hot stone and catch my breath. Then I look around.

The forest continues right up to the lake shore. Trees have crashed into the water, peat-orange shafts below water, stems washed the colour of bone above. Penetrating the underbrush is hard. Huge root plates of fallen trunks rear out of the undergrowth, the size of houses, dripping with life: mosses, gorse, willow, rowan, ferns and berries growing in the crater vacated by the tree. I tiptoe along the shore, across a beach of fine red sand unmarked except for the softest imprint of a three-toed wading bird. This is an island without humans but home to fourteen species of dragonfly; it is a little frightening, wild nature almost haunting in its indifference.

The birds are having a party, their song rich and varied and dense. Birds of different sizes and colours hop and dart through the canopy. A brown wader eyes me suspiciously from the safety of a submerged log as I pick my way along the shore around the boulders crusted with lichen rust red and coral green. Xylem and phloem, the fibres of trees, are very long in old-growth forests, making the resonance of sound better. It appears that birds can tell the difference, and their songs literally chime with their surroundings. The forest echoes their calls: yes, this is a good place to find food, to build a nest, to raise young. And studies have shown that birds lay stronger, larger eggs in older forests.

Twisted pines grow in the most unlikely of cracks in the rocks. Dead trees, standing and fallen, are everywhere. This is the signature characteristic of wildwood – dead trees are allowed to rest where they fall. Dead trees support far more life than living ones, hence the density of bird life. Some species like tree pipits and redstarts associate only with old-growth forests because of the volume and species of insects. The great spotted woodpecker nests only in dead Scots pines. Even more niche, the pine hoverfly breeds exclusively in wet hollows of dead Scots pines. No wonder it is almost extinct in Scotland.

In any carbon cycle, death is the engine of life. When a tree dies, wood-boring beetles enter the sapwood and begin the process of decay. Then fungi enter the spaces along with wasps, spiders and other insects, and invite other fungi. And in the final stages, the humification phase, soil organisms convert the last of the wood molecule lignin to soil. The cycle is complete. Because of its high resin content, a mature Scots pine takes forty years to decay, releasing nitrogen slowly into the soil, feeding the grubs and bacteria that are the bottom of the food chain for insects and birds.

The soil on the island is light brown, almost red, and fibrous to the touch; beneath the crumbling deadwood is a skin almost, a fabric of roots that would challenge any spade. A living tree is 5 per cent living cells by volume, a dead one 40 per cent, and in a virgin old-growth forest up to 40 per cent of the entire biomass might be dead, supporting far more life; the insect load in virgin, unmanaged forests is exponentially higher. Total regeneration of ancient woodland, the goal of many of the conservation efforts under way now, cannot happen until the trees that are currently young have died and rotted away. For many of the fragments of Caledonian pine wood only recently allowed to regenerate, this is 400 or 500 years from now. There is a subcategory in conservation classification beyond 'old growth' called 'true old growth'. True old growth has soil structure and a complex understorey (the layer of vegetation beneath the main canopy of a wood) that can only come from the accumulated deaths of generation after generation of tree. This is the significance of the islands of Loch Maree.

A huge black and yellow striped dragonfly inspects my face before disappearing over the bronze surface of the lake. Further off, strings of horsehair lichen hang limpid in the still evening air, out of reach in the impenetrable forest. Try as I might, in bare feet I cannot get very far into the thickets of the wood. The pines are jostling for every inch of space on the rock. The island looks as if it might sink under the weight of trees. I wade back into the loch and tread water facing the island. The amber liquid slides around my limbs and into my mouth. It tastes sweet; the trees have filtered it. Why, I wonder, have the sea trout and the salmon stopped returning to Loch Maree for the last three years – has the water become too warm?

From a distance, the canopy of pines has the appearance of a carpet, a single organism made of five-sided cells. The bark is a kaleidoscope of colours, from grey to orange to scarlet. A very old stunted granny has encased one of the boulders next to the lake in roots, her craggy arms extending in crooked lines out over the water.

As I return to the shore, the last of the sun paints a stripe across the canopy receding in golden splashes behind me. A black-throated diver skims low over the water, flashing its white undercarriage a short distance from my face. I shake off copper drops onto the pebbles, teeth chattering, while the sky slowly drains of colour and a cuckoo sounds in the forest. The call of this once common so recently rare bird does not inspire rejoicing. Instead I hear a plea. Look! See! Endangered species all around you in a lake that has lost its fish. Climate breakdown demands that we keep looking at our surroundings afresh. I sense that the struggle to overcome the gap between our idea of the world and the reality will be a feature of the years to come; our imaginations will always be playing catch up. I watch the light deepen and the lake flash magenta before reverting to obsidian black. For a long time the pink clouds echo like a reproach while the blue night fails to fully arrive, suspending judgement.

Glen Feshie, Scotland
57° 11' 40" N

If Loch Maree is one version of the past, then Glen Feshie is its future echo. To get there I drive across the continental divide, the gash of Loch Ness that slices the Highlands in two, climbing up the road from Inverness to the Cairngorms, into the heart of the eastern population of pine woods seeded from Ukraine. Glen Feshie is one of the magnificent valleys on the north-west side of the massif where the forest has been released from the tyranny of grouse and deer. This is the place, I've heard, to look for the natural treeline in Scotland.

I arrive in the evening, the day before midsummer, and pitch my tent by the river. The brown water is dark in the depths under the bridge, and cold. I find a flat spot by a ditch that seems to have been dug for drainage; no inch of this fabled glen is unplanned. In the still-bright sunlight I walk up the valley and come to a spot where the path widens and a vista of sheer grey hills opens out. This was the setting for *The Monarch of the Glen*, a famous painting by Landseer of a princely twelve-point stag framed by the crags above the valley. Familiar from whisky bottles and Highland kitsch, the image romanticises a landscape devoid of indigenous crofters or trees, devoted to deer and the Victorian love of hunting.

The view looks very different now. The river still sprawls through the depopulated valley, rushing in unruly shallow channels over pale pink and yellow balls of granite; the tops of the moor are still mottled brown and purple with heather, but the valley bottom is a feast of evergreen, and the massed ranks of pines are storming up the side of the hills to find their natural limit. They have been set free. Glen Feshie is an attempt at a new approach to land management: re-wilding.

The history of this land can be read in the hillside. There are the straight lines and even heights of the former plantations slowly being let go, then the softer, gentler canopies of the ancient stands that have persisted and the thrusting spiky tops of the newcomers, offspring germinated since the re-initiation of the natural forest fifteen years ago: here solitary, here clumped together, but everywhere blooming in a riot of iridescent green.

The ground off the path is as deep and springy as a mattress, full of moss, heather, blaeberry, grasses and tiny flowers. Granny pines are ringed with nurseries just beyond the reach of their canopies; they cannot regenerate under them. This means the Caledonian forest was probably a dynamic, mobile wood, shifting with every generation of trees. Venturing upstream, huge twenty-, thirty-, forty-metre trees command the bends in the river, supported on all sides by seedlings rising to five metres. At the head of the valley the glen divides into a spectacular view, a bowl on either side scooped out of the earth by glaciers.

A hare crosses the path. A crested tit is making a racket, and three white horses graze unconcerned in the last of the evening sun. It is an awe-inspiring vision of nature taking over, an example of what can happen when trees are allowed their own dominion, or perhaps an example of what once was – minus the indigenous crofters. Flies, moths, sap, honeydew, parasites and fungus swarm over the old dead trunks, which lie where they fell. Birch, rowan and alder are mixed in, cooperating with the pines in what looks like a social scene. Everything is shining, blooming, buzzing. As though nature has burst into song. Indeed, in the morning I wake to the sound of a capercaillie calling in the mist – *cuttuck, cuttuck* – sounding like a horse's trot. The Celts called it the 'horse of the woods'. There's not a deer in sight; they are their proper, elusive, self.

'Deer are not the enemy of the forest,' says Thomas MacDonnell, the factor of Glen Feshie, when I meet him later; they are the foresters, they keep the grass down and encourage herbs. Capercaillie in particular follow patterns of deer browsing within the forest – if there is a forest.

The roots of the word wilderness in English are 'wild deer place', but this romantic ideal was allowed to go too far once the High-lands were cleared of people. In a sense, an excess of re-wilding in favour of deer tipped the ecosystem out of balance. During the deer-stalking centuries, the 1800s and 1900s, there were fifty deer per square kilometre in Glen Feshie. Now there are one or two. And the critically endangered capercaillie are coming back. Breeding males have just established a new territory, or lek, at Glen Feshie, a badge of pride for everyone concerned. A lek, a kind of gladiatorial clear-ing in the forest for displays by male birds, is only possible in a large volume of old forest. Capercaillie rely on a diet of blaeberries, which only grow in the dappled shade of well-structured pine woods.

Glen Feshie is the jewel in the crown of Wildland Ltd, the pri-vate property empire of Danish businessman Anders Povlson, which is dedicated to re-wilding. Wildland was a pioneer in a movement that is now gaining ground. It is universally acknow-ledged that industrial agriculture and urbanisation have helped annihilate over 40 per cent of British wildlife during my lifetime (I was born in 1974), depleting the soil to dangerous levels. 'Nature recovery' has become a mantra, if not a government priority, across Europe, and political parties now bid to plant more trees than their rivals. Re-wilding is both trendy and emotive. Some in the countryside advocate it passionately, others see it as a mortal threat to their culture and history, to an entire way of life.

It seems strange that trees should evoke such extreme reactions, but Glen Feshie poses a fundamental question about land. Without sporting income or productive value as forestry or for commercial agriculture – without the prospect of a financial return – what is land actually for? The simple answer is life. We need land to grow food, but we also need to set aside enough wild land to produce the oxygen and the biodiversity that we require to survive. There is enough land to feed everyone if it is distributed and managed prop-erly and if issues of lifestyle, of consumption, of values, of equity and justice, of the disconnect between livelihoods and the living world are addressed. If there is not enough diversity and abundance in the living world, then there is no life, human or non-human, at all.

*

Thomas MacDonnell is an unlikely revolutionary. In green fleece and hiking trousers, with clean-shaven cheeks and close-cropped silver hair, he has the look of a rambler about to step out for a walk, but his penetrating black eyes burn with a fire that I have previously encountered only in priests and politicians. He settles into a wing-back chair with a view over a housing estate in the town of Aviemore, where Wildland has its offices, and turns his poker face towards me.

Thomas MacDonnell is probably responsible for the death of more deer than anyone else in Scotland. As the conservation manager of Wildland, for over fifteen years it has been his mission to get grazing levels in Glen Feshie down to a number that will allow the trees to regenerate. The thriving forest is a living monument to his efforts. With the success of the programme finally visible on a landscape scale, the tide is beginning to turn in his favour, but it has not been an easy journey.

Close friends from childhood accused him of endangering their jobs. In packed village halls he was shouted down when he tried to explain the rationale behind the deer cull and his 200-year vision for Glen Feshie. Farmers, deer stalkers, ghillies and gamekeepers were anxious about the impact his plans would have on their jobs, their culture. For Thomas, emotion had nothing to do with it. He was trained as an engineer. He has always wanted to understand how something works, analyse what's wrong and then devise a solution.

'In a way they were still trapped in a colonial way of thinking,' says Thomas. He assesses the criticism that he faced dispassionately. The land, he says, was 'burned out' literally and figuratively, and the communities tied to patterns of Victorian land use are 'refugees from the nineteenth century'.

When he became the factor – land manager – of Glen Feshie Estate twenty years ago, it was clear to him what was wrong. What this was had been clear to governments and landowners since the Second World War, in fact. Government commission after commission had tried to reduce deer numbers but been unable to persuade or unwilling to enforce a cull on landowners wedded to the income from deer stalking. As a young man Thomas had spent many wet and cold days fencing timber plantations to keep deer out. He had grown up in Glen Feshie and the neighbouring valley. He understood the mechanics of the ecosystem and was curious about what would happen if the recommendations of the deer commissions were actually implemented. And when the estate changed hands, Thomas acquired a new boss who was open to his ideas about doing things differently. This was his chance to conduct an experiment.

'Sometimes, the most radical and brave thing to do is to do nothing,' he says. Although culling deer is not exactly nothing, he means not managing the land, and this goes against hundreds if not thousands of years of practice, including the shifting nomadism of the crofters, which in its way was management too. It needed courage. There was a moment, a spiritual moment, he recalls, in 2006. He had shot 5000 deer off Glen Feshie Estate in the preceding three years in the face of intense criticism – deer are not fenced in one particular estate, but free to roam across the Highlands. There had been three years of low deer numbers but the pines seemed unwilling to come back.

'They were dark days. *Bloody hell*, I thought, *perhaps I'm wrong.*'

Then, in June of the third year, Thomas was walking in Glen Feshie when he stopped by a familiar granny pine that he had long held to be senescent. All around the old tree were tiny green fingers poking up out of the grass – a ring of seedlings. He looked up into the canopy and saw the most fantastic seed array – the granny was not sleeping at all.

'It was as if someone had flicked a switch. They just came. I almost cried! Maybe they had realised someone was trying to help them.'

Scots pines do not produce a good crop of cones every year. At higher latitudes harsh winters and short growing seasons mean that sometimes a tree will not flower and produce seed for three or four or even a dozen years. The succession of climate is important – what happens one year influences what happens the next. A warm sunny summer will usually yield abundant flowers the following year, and a tree carries three years' worth of cones at any one time.

The bud of needles that appears in the spring unfolds into a resiny brown crop of flowers along the stem. These tightly packed bombs carry the male gamete. These begin life looking like miniature cones but soon crumble into powder to be dispersed on the wind. In May and June clouds of yellow pollen can be seen wafting through a mature pine forest. In a healthy ecosystem this can form a scum on the surface of ponds and lakes – the pollen has two air bladders specifically designed to keep it afloat. Not only do Scots pines give their pollen the best possible start in life, they also synchronise flowering with other pines to increase the chances of successful fertilisation. Scots pines have been shown to be synchronous in their flowering across distances of up to 200 miles. How do they know? Ecologists have suggested hormonal communication on the wind, as well as via fungal networks underground, or it could be a deeply embedded genetic trigger activated by certain climatic thresholds. But no one really knows, yet.

The female cone starts off as a tiny purple boil atop a young stem. The pollen grains land on the red-purple surface of the cone scales. The sugar of the pollen and the moisture of the cone's resin interact to form a sweet pollination droplet which is sucked into the micropyle tube – basically a canal into the cell, where it makes contact with the nucellus. The cones harden further over the summer and grow to under a centimetre before pausing for the winter. It is only when the pollen tubes resume growing the following spring that fertilisation occurs as the extending tubes penetrate the nucellus. They reach their full size by the end of the second summer, but they are still green and sticky, only browning and hardening through the autumn and winter. The third spring is when the cones crack open as the scales spring back to reveal the seeds and shake them loose into the wind.

A Scots pine seed looks a bit like the wing of an insect, with a hard seed case and a long papery wing acting like a sail to catch the breeze. The seed endosperm crop is an essential food source for crossbills, siskins, tits, woodpeckers and red squirrels. Red squirrels shred the scales to get at the seeds, eating up to 200 cones a day. A hectare of trees is needed just to get one squirrel through the winter.

The crossbills, by contrast, prise open the scales with their crossed mandibles before sliding their hungry tongues inside to pull out the seed. Ridges on the bird's palate allow it to de-husk the seed as it is swallowed. Rodents and insects love pine seeds so a forest must produce a seed crop larger than the appetites of the rest of the food chain if it wants its seedlings to stand a chance. This seems to be the reason behind synchronous mast years – when all the trees produce a spectacular crop of seed at the same time.

Mast years, though, are becoming more common, and warming is confusing the trees, triggering the pines to release their seeds earlier and earlier and threatening to disrupt the normal seasonal cycles of the forest. At the moment there is still an overlap of seeds in one cohort of cones and the availability of seeds in the following cohort of developing cones. But if the generations of cones on the tree become entirely decoupled from each other, a whole raft of seed-dependent species could go hungry. It is not just the animals that eat seeds who will suffer; the pine flowers rich in nectar are key food sources for dozens of species of moth, butterfly and other winged insects. The larvae of moths spend the winter tucked in the protective folds of Scots pine bark, to emerge from the chrysalis in the spring ready to feed on the flowers. But if the tree has already flowered, a whole generation of moths will die. If the moths die, birds will have less to eat, and so on. Twenty-one days is the maximum window of divergence from the normal flowering date that most species of caterpillar and moth larvae can tolerate. In 2020 we are currently at eleven or twelve days. The most worrying phenomenon of all for the pines would be a decoupling in the timing of flowering and pollen release. If synchronous flowering depends on learned genetic responses to climatic signals and not on communication under the ground or through the air, then asymmetric warming – uneven weather patterns in which some parts of the forest experience a different climate than others – could mean that seeds might not form at all.

Even if the seeds form and then avoid detection and consumption by rodents, birds and beetles, germination is not assured. Seeds need exposed gravelly sites or areas with shallow peat like road verges and riverbanks. Heavy peat is poor in nutrients and unlikely to hold the mycorrhizal fungi that pine need to access phosphates and nitrates in otherwise nutrient-deficient soil. Building and extending the mycorrhizal relationships from existing forests is by far the easiest and most reliable way to ensure successful germination. Even where there hasn't been forest for generations, the mycorrhiza are likely to still be present in some dormant form.

The soil is our main guide to where planting or encouraging trees stands a chance, says Thomas. Wherever there are mushrooms, ferns, bracken and particular kinds of woodland plants like violets there was once forest. Rings of mushrooms are usually the outline, the long-ago earthwork of a tree stump. There are between fifteen and nineteen ecto-mycorrhizal fungi (fungi growing around the roots) in a mature pine forest, and they play a role in everything from carbon and nutrient transport to lichen cover, taking sugar from the tree and providing it with minerals in exchange. Planting trees without regard for the essential symbiotic 'other half' of the forest below ground may be far less effective than allowing the ground to evolve into woodland at its own pace. Oliver Rackham describes a planted oak wood in Essex that even after 750 years still does not possess the orchids, plants and mushrooms that you would expect of a natural wood.[12]

Wildland now has fifteen years of baseline data to inform its future management. It employs ecologists alongside volunteers to survey the ground in seven-metre quadrants to understand how ecological succession takes place. The ecologists count the dung to establish deer numbers, they measure the height of the heather to see if it is shading out other plants, and they count the pupae of the winter moths on the leaves of the blaeberry plants which the capercaillie feed to their chicks in winter.

Thomas still does not know what will happen, how the different species will respond – the timescales involved are immense and hard for a human mind to grasp – but he is enjoying watching and measuring the change. Suddenly, pines are everywhere, and not just in Glen Feshie. In the early years it was 200 stems per hectare, now it is 6000. The trees are gathering strength.

The first goal is habitat restoration. Then Thomas will try to find a metric for what is a manageable number of deer and make the wild vision pay for itself through tourism and other 'natural capital' initiatives. But, in the absence of an apex predator like the wolf, Thomas will still need to shoot deer, and he will still need to work with Glen Feshie's neighbours since the deer are wild and the mountains are impossible to fence in their entirety. This is the thinking behind the establishment of Cairngorms Connect – a joint venture between Wildland and the neighbouring estate of Abernethy run by the Royal Society for the Protection of Birds (RSPB), the Forestry Commission Scotland, Scottish Natural Heritage and the Cairngorms National Park, a stunning 600 square kilometres of land unmanaged for its wilderness potential.

Thomas's parting shot, before inviting me to a midsummer ceilidh – an evening of Gaelic singing and dancing – in the bothy at Glen Feshie that evening, is a hint at the scale of his ambition. Half of the estates in Scotland are for sale. Wildland is busy acquiring several of them. By the end of 2020 Wildland will own 221,000 acres, making it the largest landowner in Scotland. Glen Feshie, he tells me, is 'only' 43,000 acres.

'I'm gardening here,' he says with an ironic chuckle. It is the only time in our meeting that I actually see him smile.

Further out of Aviemore, in a converted shed adjacent to the car park of Scottish Natural Heritage, is the office of Cairngorms Connect. The humble location belies the ambition of the partnership as a maverick outlier in an otherwise still very conservative conservation industry. The crisis in the natural world requires confronting deeply held assumptions and habits, even among those paid to look after it.

Mark Hancock is a scientist seconded to Cairngorms Connect by the RSPB. He makes me a cup of tea which we take outside into the sunshine to a tree stump in a field beside the office hut. The whole breadth of the Cairngorms massif – granite and gneiss rising to over 1200 metres – sparkles on the horizon, with Glen Feshie on its western end and Abernethy Forest on the east. Cairngorms

Connect is aiming to establish a corridor of wilderness between the two.

Mark pushes his glasses up the bridge of his nose and chooses his words with scientific caution and precision. I get the feeling the scale and aims of the plan are unfamiliar territory for him and others used to the language of science, of control, of management. He readily admits that Cairngorms Connect has turned convention on its head, going 'against the grain of a hundred years of conservation history, the practice of hands-on management'. But if it works, and Glen Feshie suggests that it will, then it could transform the way British conservation thinks and acts on a national scale. It could change how we think about the value of the land, even change ideas about the purpose of national parks.

Britain is desperately trying to increase the size of its woodlands. At 13 per cent forest cover, the UK is far behind the European average of 37 per cent and the global average of 30 per cent. Moreover, much of that 13 per cent is plantation – a single crop grown for timber – with very little diversity of species living among the trees or beneath them. Indeed, plantations are not really forests at all.

Sweden and Finland account for a third of the forests in the European Union with 68 and 71 per cent of their territories wooded respectively.[13] Even then, Finland's forests only absorb about half of its greenhouse gas emissions. For the UK to triple its forest cover to the European average would be a revolutionary proposition, but still not enough. Forests could soak up a quarter to a third of global human carbon dioxide pollution, but the scale of the land use change needed to achieve such a goal is mind-boggling: studies cite an area the size of India, equivalent to the land given over to global agriculture already, without accounting for the trees lost each year. Fifteen billion trees – thirty million hectares of forest – are cut down every year around the globe, and about the same amount is lost to wildfires. Cairngorms Connect is a tiny glimpse of what could happen if carbon sequestration or woodland creation became national security issues, which they soon might. Companies eager to offset their emissions are already pushing up the price of large Scottish estates.

At present, though, Cairngorms Connect is not a military operation but a fragile experiment in understanding nature as a dynamic work in progress, in thinking of forests as mobile communities and land as constantly in flux. Mark's chief interest is birds and their habitat. In this respect he is less concerned about the volume of land represented by the partnership but about its topography; the diversity of ecosystems that it could support. Could the forest migrate if it had to? Is there an impediment to it moving upslope if it wanted? And when it got there, is there enough cold mountain left?

In an undisturbed system, the pines would give way at the natural treeline to montane woodland, what the Norwegians call 'the willow zone'. The montane system of willow and rowan prepares the ground for the pines over time, forming the first phase of the forest. But deer love willow and rowan best of all. Some species of willow have been grazed so heavily that only one or two specimens remain in the wild in Scotland. The montane zone of delicate shrubs and berries that continues beyond the limit of continuous forest is the biggest casualty of overgrazing by deer in the Highlands. It is all gone.

Ben Lawers, a National Trust for Scotland reserve a few hours south, has the best collection of montane species in the UK, a kind of seed bank for the rarest habitat in the country. It comprises a few acres of a south-facing enclosure protected by a high and secure deer fence with spring-loaded gates. When I visited on the way to Fort William, I found neat circles clipped out of leaves on nearly every other plant, marking the presence of caterpillars tight in their furred caskets on the hairy underside of the leaf. Butterflies bobbed all over the meadow and the number of birds per square metre was astonishing. The scarcity of the habitat makes it a refuge for species desperately searching for their traditional food. It had the feeling of a zoo almost; a tiny scrap of botanical garden mistakenly located halfway up a vast forbidding mountain, protected by fortifications. There is a long way to go before these very rare species will be safe again in the wild.

The willow zone is a critical habitat for particular species of birds and insects. This is the place of 'edge effects'. As the forest

reaches the limit of its growing range, different species give up at
different heights, creating a rich zone of diverse species combinations. More light and greater temperature ranges create different
kinds of transition zones. It only takes the addition or subtraction
of one key species for the whole balance of life to shift. Without
pines, willows can reign. And beneath the willows, hair grass and
feather moss that need their protection close to the ground can
thrive. But they are also the first things to go when grazing gets out
of control. Rare lichens abound here, forming associations with
the plants of the willow zone, like those that grow within heather
or the woolly fringe moss *Racomitrium lanuginosum*, a lifesaver
for insects in winter. It can survive for twelve months without
water by absorbing atmospheric moisture. In deep snow the woolly
fringe forms a cushion with its own microclimate up to twenty
degrees warmer than the surrounding snowpack. If you lift the
woolly lid you will discover thousands of insects packed inside,
keeping warm.

Insect habitats are of course good for birds. Abrupt landscape
shifts – the straight line of a pine plantation – are unnatural and
unhelpful. Birds like mosaics, with different insects, seeds and berries available at different times. Now, within the Cairngorms
Connect area, transitions are becoming blended as forest and moor
are once again blurring. These are the marginal habitats that create
the conditions for diversity. The ring ouzel, for example, a member
of the thrush family, is a treeline species, moving between habitats.
Dotterels, snow buntings and hen harriers are all moorland birds
that occasionally need the comfort, or the food, of the forest. Hancock fears these species might be the 'extinction debt' that we have
yet to pay – whose habitats have been so degraded, whose numbers have been so diminished beyond their ability to recover.

A key aim of Cairngorms Connect is to allow trees to re-establish
up to the natural treeline. This can then guide other efforts, so
montane woodland restoration projects will know where to start.
A body called the Montane Woodland Action Group is busy planting willows and juniper at 600 metres in some venues, but it is
delicate and fragile guesswork. And the isotherm, the theoretical

growing limit, is not static. Those willows might be in the middle
of a pine forest in a few years' time.

'No one really knows where the treeline is, or should be, in Scotland,' says Mark.

The conventional wisdom is that there is only one area where
the trees have hung on at their natural limit, where you can see a
proper gradation of krummholz – trees progressively stunted by
altitude and exposure. It is a place the writer Jim Crumley calls 'a
woodland shrine', Creag Fhiaclach.

The next morning I set out early up the path to Creag Fhiaclach
('toothed crag' in Gaelic), which begins where Glen Feshie opens
out into the valley of the River Spey. Here the Forestry Commission plantation of Inshriach Forest stands like an intruder between
Kingussie ('head of the pine forest') and the well-preserved pine
wood of Rothiemurchus ('wide plain of the firs') around the picturesque lake of Loch an Eilein.

I make my way out of the closed canopy of the plantation, with
its gravel roads, straight lines and eerie silence, into the scattered
open woodland of a bog where birds are suddenly all around me. I
clamber over hummocks waist high as my boots disappear into soft
cushions of moss. A stream clear as glass tinkles quietly between
rich banks of peat topped with grass. Heather, ferns, cloudberry,
rushes and sedge grow along the riverbank, and hard black domes
of peat rise above it from which sprout ancient bog pines, gnarled
and twisted, their layered branches sculpted by the prevailing wind.

I pick my way through dense undergrowth of heather and gorse,
the whip of saplings of rowan, hazel and alder and the ubiquitous
scratch of pine. The words of the bog-pine dendrochronologist
Rob Wilson come back to me: 'Always remember, pine is a weed,
given half the chance.'

A small brown bird follows me at a distance going *pee-wit*. I
come across the feathers of an owl by a burrow and then a large
cracked white egg the size of a grapefruit beneath a flat-topped pine
tree. Looking up, I see the jumbled sticks of a golden eagle's eyrie
just visible on the crown. There are droppings of rabbits. Newts in a

pond. Deer have recently visited; a rowan, three metres high, has no leaves or branches from the waist down. I feel like this is the boreal forest as intended. Without, of course, the bears, lynx and wolves that once were here. It is older than Glen Feshie, without the thrusting newness of a generation of upstart pines. The re-wilded valley had a feeling of joyful surprise, of exultation in freedom, whereas the more established rhythms here speak of a settled pattern of life.

Ascending towards the sound of water, I finally come across a path trodden into the black, mineral soil. The wood is spread out: ancient trees stand at intervals with the understorey filling the gaps. Giant hummocks are an ecosystem unto themselves. There is more birch than I had expected, venerable wizened creatures, their exposed red roots trailing in the stream, with skin like ancient lizards, ribbed and cracked. The stream is full, clear and loud, roaring among the trees and misting the banks of moss and ferns.

Tight fists of bracken are beginning to unfurl, and tiny red hairs of feather moss tickle the air. Mosses of all sorts are everywhere: sphagnum, club, feather and antler, pushing out their little spores on fragile translucent stems. They hold the moisture of the forest, a hyaline chemical allowing them to retain one thousand times their weight in water. The moss is the first sponge, catching the rain and releasing it slowly according to the capacity of trees and soil to absorb it. In the process it fixes nitrogen, fertilising the forest.

The dominant species is sphagnum – peat moss. The germ of coal, and of diamonds. This is one of the winners from global warming – sphagnum loves carbon dioxide. But too much of it can smother the understorey, changing the successional pattern of the forest. In Siberia rampant moss is already hindering the establishment of larch seedlings. The accumulation of carbon dioxide acidifies the soil, in the same way that CO_2 acidifies the ocean, suffocating other plants.

In the Carboniferous period, when the atmosphere had a lot less oxygen than now, mosses and ferns like horsetails were dozens of feet high, monstrous triffid-like plants that gorged themselves on carbon dioxide. The triumph of the gymnosperms and angiosperms – coniferous and deciduous woody plants that slowly oxygenated the atmosphere – cut the equiseta, ferns and mosses down to size. With current warming, we can expect them to rise again, with serious consequences for forestry and agriculture as acidic air empowers the moss, choking fields.

The path is unforgiving, up towards the open moor now visible above and across the stream, which narrows to a gully. The north side of the gully is still forested, facing south. On the opposite side, the trees abruptly give way. Fifty metres further up on my side of the stream, the krummholz begins to appear and then rapidly falls away. I count the bracts, the notches marking the yearly growth, of the last pine tree beside the path. It only reaches my waist but it is sixteen years old. Further on, a whole miniature birch tree the size of my hand is at least ten years old. Juniper and willow grow flat against the hill now. In a hollow up above I spy a rowan and more upright willow. Here are the tentative beginnings – or perhaps remains – of Britain's only treeline transition zone.

Before I visited, I had no idea this willow zone habitat was so rare or so important. This is where the ring ouzel, the hen harrier, the snow bunting, willow warbler and lesser redpoll belong. It is where the redwing, lapland bunting and bluethroat could be, if the habitat were restored. So much history lies behind the glib and anodyne phrase 'habitat loss' that it is often hard to grasp what is meant and, therefore, what we must do differently to reverse the damage. The birds need the delicate balance between lichen, moss and insects, which in turn relies on the layered protection of a few species of willow, some of which are rarer in the British Isles than pandas are in China. I thought I was paying attention, but a whole different level of noticing is required.

The palette of the hill is red with moss, pink from the granite, bright green with shoots of blaeberry, the orange and red and white of lichen, all set against the blanket grey of heavy rainclouds pressing down. Below me the forest is an unending green, punctuated by trucks on the highway and the iron grey of the River Spey looping through the trees. Far off are the bare burned-brown grouse moors and the geometric blocks of plantations that now appear to me to have all the implied violence of brutalist concrete architecture.

I reach the summit of Creag Dhubh (756 metres – 'black crag') just as the sideways rain comes biting. I turn downhill at the Argyll Stone. A little way to the east lies Carn a'Phris-ghiubhais – 'hill in the thicket of pines', suggesting that this barren hillside was not always so windswept. And indeed in between the cold sheets of water lashing in from the east, I look down the length of the opposing ridge of Creag Follais and what do I see, illumined in the shifting grey light of the storm? The unmistakable outline of pine trees, well established by now, bending precociously in the wind, laughing at their own audacity, way above their brethren lower down.

The bothy at Ruigh-aiteachain, one of a network of mountain huts made available to hillwalkers for free, lies at the head of Glen Feshie at the spectacular confluence of the River Feshie and the Lorgaidh burn, where the glen divides into its iconic glacial bowls. Descending from Creag Fhiaclach, it's a long walk the length of the glen through a patchwork of landscapes: heather and swamp grass beside the gravelly riverbed, clear forest ponds sticky with insects and the tall, rhythmic landscape of an old pine plantation.

This being Wildland's Glen Feshie, the plantation is being let go and encouraged to re-initiate natural processes; thus, here and there fallen logs are left to rot, some thinning has taken place and in clearings saplings are thriving. But the structure of the plantation is monotonous: all the trees are the same age, the trunks dead straight competing for the light where they have been planted deliberately close together. Thinning has allowed light to reach the forest floor, though, and so now the ground between the trees is a lurid green of blaeberry, heather and feather moss. Too much sunlight and the heather takes over, too little and the blaeberry cannot grow. The world of the understorey is a precise calibration of light, nutrients and mycorrhizae that has a direct relationship to the density of the canopy, finely controlling the conditions for what is able to live below. This is why the capercaillie, and other birds, have come back.

On a rise beside the river is Glenfeshie Lodge, where guests pay a lot of money to have a 'wild' experience, rather like an African safari. Further on, hidden in a clearing of ancient pines, the stone bothy built and maintained for less wealthy tourists has two rough-planked rooms heated by Norwegian cast-iron stoves. Outside is a composting toilet and a bucket beside a cool clear stream.

The midsummer ceilidh mentioned by Thomas is hardly a dancing party; there is barely room to stand up in the tiny room, baking hot from the roaring wood stove. About twenty people have come to listen to a Gaelic legend sing songs about the old ways.

Margaret Bennett is a singer who has written several books, including *Scottish Customs from the Cradle to the Grave*. Before she sings, she lays her stick down, takes off her thick glasses and settles herself beside me in front of the stove. Margaret wants to talk about the magic of trees. Thomas has told her I am writing about the wood. Her black and silver hair is tied in a simple knot and her blue eyes appraise the room as she talks.

She tells me how in springtime girls used to wash their hair in buds of birch, and go to church smelling of the tree, how her mother planted a rowan outside their house for luck; it stands there still. We talk of how pine has always been medicinal, its needles used traditionally for fumigating homes, for respiratory conditions and now in industrial applications like camphor, insecticides, solvents and perfume. The Native American Cayuga peoples boiled the pith of pine knots to release the antibiotic pinosylvin, the same chemical used in antibiotic salves to treat skin conditions and bites. Margaret talks of the functioning clan system, where no one had a title deed to any land; the forest and the hills were maintained for the benefit of all. She rues how the monarchs granted land titles to clan chiefs instead of the collective, paving the way for the land to be bought and sold. This is the background to her songs: the people of the Highlands, removed to make way for sheep and deer and still absent from the hills. The modern image of the glens is intimately tied up with the history of exploitation, clearance and militarism directed from south of the border.

Later, when other guests have arrived and we have eaten several kilos of venison burgers, Margaret sings songs of young girls waiting for boys gone away with the Royal Navy or the Black Watch,

tales of Captain Macleod and his phantom piper. There are others
about the redshank and about a handsome drover with 'calves like
a salmon' taking Highland cattle across the hills to Crieff with
rowan branches plaited in their tails for luck. Thomas's brother
Sandy accompanies Margaret on the pipes, and the chief ecologist
of the Cairngorms National Park, a tall dark-haired man named
Will, joins in with his banjo. Margaret has the twenty of us singing
along in Gaelic, and there are tears in many eyes by the end.

'So, you see, we haven't quite forgotten the old ways yet,' says
Margaret, her irises flashing the same colour as the half-night of
midsummer midnight outside.

In the morning I'm up early with the milky dawn. The moon still
hangs above the shoulder of the mountain to the south while the
sun burns behind it. It is the summer solstice. Inside, twenty bodies
lie in sleeping bags on the floor. Over breakfast, I explain to Will
my presence at this curious impromptu gathering. I tell him of my
pilgrimage the previous day to Creag Fhiaclach and the treeline,
and he smiles to himself and nods his head knowingly.

'That's not the treeline,' he says, pulling out his phone.

Earlier that week Will had just finished a late night at the office
drafting the forest strategy for the national park, and, as he likes
to do to relax, he set off up the hill to camp out for the night at a
place he hadn't visited before. He pitched his tent in the dark, and
when he woke up and poked his head out, he realised he was still
in the forest. There were lots of tiny trees dotted about. He was
high on the Cairngorm plateau. He shows me a photo of a well-
established tree with many stems growing among bog grass and
heather, with an altimeter in the frame. It reads 1045 metres. He is
not surprised that I saw pine trees climbing up the slope even
higher than Creag Fhiaclach. In all likelihood, the natural treeline
in Scotland could already be above the summits of its highest
mountains, he says. And if it isn't yet, it soon will be. In Sweden,
where an intact pine treeline has been monitored for decades, it
has moved 200 metres upslope since the 1960s.[14] And recently,
when summer drought killed the leading edge of the birch in

Sweden, the more resilient Scots pine took over at the vanguard of the treeline, leapfrogging the birch.[15]

Cairn gorm is Gaelic for 'blue mountains'. Apart from several months between June and October, snow covers the plateau. From a distance, this can give the ridges a tinge of blue. But in old Gaelic *gorm* denotes a more greeny-blue.

Irish-born Canadian botanist and chemist Diana Beresford-Kroeger has written extensively about the boreal forest. She told me that the colour of pine needles is of critical importance to their resilience and ability to adapt. Needles that are bluey-green have a cuticular surface that is two or three microns thicker. The waxy layer attenuates less light, so the chlorophyll of the needles' cells appears bluish. This extra thickness protects against extreme heat or extreme cold; it holds the key to the epigenetic ability of the pine to be more flexible with climate. When collecting specimens, Diana advises, one should always seek the bluest conifers from the highest elevations as the seed source for the future. These will be the hardiest, she says; a difference in hue could buy a species 'hundreds of years'. Trees at altitude tend to evolve this trait more readily, which is why high mountains covered in conifers often look blue. Were Scotland's 'wooded heights' once these very *cairn gorm*s? And could they be again?

The restoration of Caledonia's lost great wood is a beautiful project and necessary if we are to re-establish our reverence for and our connection with the living world. Wildland and the re-wilding movement more broadly is succeeding in re-enchanting millions and beginning to broaden the agenda to talk about re-peopling the Highlands too. More than enchantment will be needed, however, if the new great wood is to avoid becoming a zombie forest.

The United Kingdom Forestry Commission forecasts that with increased carbon dioxide, warmer temperatures and a longer growing season, Scots pines will do well in Britain this century.[16] If humans allow them, Scotland's pines do indeed look set to turn the mountains blue once more. But perhaps not for long.

Scots pine currently inhabit a wide climate niche from the

treeline to southern Europe. While Scotland is at the northerly
limit of that range at present, the territorial window for Scots pine
is shifting north. Further south in Europe, drought and heat stress
is already causing pine needles to turn brown early and crumble.
Research from 2008 predicted that one to four degrees Celsius of
warming would reduce the survival of Scots pines at latitudes
below 62°N. The Cairngorms are at 57°N.[17] This stark picture is a
grim illumination of what climate breakdown means for trees and
forests.

Global warming has the same effect as inexorably travelling
south. The UK's current climate velocity is the equivalent of moving south at around twelve miles per year. By 2050, London is
projected to have a climate similar to that of Barcelona.[18] While that
might sound nice for some humans, for Scots pine it spells trouble.
Two separate models – devised by the UK Met Office and the
University of Oregon – predict that on current emissions trajectories, Scots pine could disappear from lowland Europe, including
Scotland, by the end of the century. The Met Office model goes
further, predicting survival of Scots pines only in northern Fennoscandia, Russia and the Alps.[19] This echoes other findings predicting
that, even at one or two degrees of average warming, the viability
of existing forests is already starting to fray.[20] The safe space, the
climate niche of Scotland's great wood, is on the move, leaving the
trees behind.

Unless something dramatic happens, Cairngorm Connect's
admirable 200-year vision will be overtaken by global warming.
Over 8000 years of wooded history and all the birds and insects
and mammals that make up the finely balanced system of forest
that has evolved around the Scots pine could be obliterated within
the lifetime of a single tree. Where once Scotland was at the northerly treeline, above which it was too cold or too high for pines to
grow, in less than a hundred years it may find itself below the
southern limit of that range.

2. Chasing Reindeer

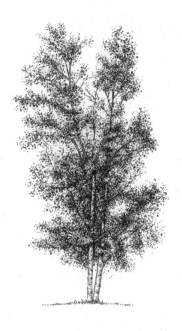

Downy birch, *Betula pubescens*

Finnmark Plateau, Norway
69° 58' 07" N

Altafjord is a wide expanse of black water ringed with great white domes of mountain. The first snow of winter has fallen during the night. The morning wind drives the sea up the funnel of the fjord towards the town, sending great heaving masses of blackness crashing into the breakwater beneath the window of the guest house where I am staying. The foam on the waves is momentary; any glimmer of light is immediately sucked back into the enveloping darkness of the Barents Sea.

Alta is the capital of Norway's Finnmark province, the crown of the horse's mane that forms Norway's jagged coastline and Europe's northern shore. I am much further north than Scotland, but here at sea level the most northerly trees in Europe are moving upslope in altitude as well as latitudinally towards the pole. The trouble is, there isn't much room for expansion. From Alta, there is nothing but water for one thousand miles until the beginning of the sea ice of the Arctic Ocean. The tundra is caught in a vice. And the people and animals that live here are trying to make sense of the rapid changes with a mixture of confusion, denial and panic.

Dawn at seventy degrees north during winter is eerie and unending. It lasts nearly the whole day. At eight in the morning a numinous lilac light illuminates the mountains to the south from behind. Thin clouds above are furred with pink, the only hint of the mighty absent sun crouching behind the horizon. It is a pre-dawn light, a kind of gloaming, but the transition never comes; the sun never rises, the day is permanently on the verge of breaking. It is disorienting. Half an hour later the sun is still beyond the rim of the world and the moon is still glowing mauve, a little lower over the roiling black sea. The steep fjord behind lurks in shadows, reminiscent of the mountains ringed with fire in the folk tales about the kingdom of Thule at the edge of the known world.

The polar night does not disrupt the routine of the modern working week. It is Monday morning, and Alta's residents are crawling out of bed, wrapping up warm, defrosting windscreens and getting into their cars to make their way through the half-light misted with exhaust fumes that do not disperse in the cold air.

On the way to city hall from the guest house, I spy children in classrooms and several queues of cars inching along the icy road on snow tyres but few pedestrians. Alta is a town built along American principles – that is to say a town built for a world in which petrol is cheap and cars are taken for granted. It is a land-scape of shopping malls, gas stations and spaced-out residential suburbs with houses on large plots. Normally at this time of year it isn't safe to be outside for long without wearing animal skins. Today it is only minus one Celsius, but the habit of driving encour-aged by urban planning is hard to break.

All along the road to the city centre are rows of young Scots pines, their orangey bark contrasting with the fresh dusting of snow and, intermingled with them, shorter, ragged-looking trees with lumpy trunks, wizened branches and fine twigs like gnarled fingers: *Betula pubescens*, downy birch. It is these trees that have brought me here, to the office of Hallgeir Strifeldt, director of planning for the municipality of Alta, at nine o'clock on a Monday morning in the middle of winter.

Downy birch is much shorter and more untidy than its graceful cousin, *Betula pendula*, silver birch, and it has evolved to survive much further north. It is one of few broadleaved deciduous trees in the Arctic and hardier even than most conifers. The down of the name is a soft coating of trichomal hairs that acts like a fur coat in the punishing cold. Often found cooperating with pines and spruce at lower latitudes and altitudes, above a certain point the birch leaves the others behind and goes on for hundreds of miles alone. Sometimes called moor birch, white birch or mountain birch, *Betula pubescens* and its variant, dwarf birch, *Betula nana*, make up the bulk of Europe's Arctic treeline from Iceland (where it is the only tree to form natural woodland) across the top of Norway, dipping into Finland and the wetlands of Karelia made famous by Sibelius, across Russia's Kola peninsula to the White Sea, beyond which the larch of Siberia picks up the baton.

It might be unprepossessing, even ugly, with its stumpy branches and pockmarked bark, but this tough little tree is a survivor and a pioneer, essential to nearly all life in the Arctic: human, animal and vegetative alike. Used by humans for tools, houses, fuel, food and medicine, it is home to microbes, fungi and insects central to the food chain and is critical for sheltering other plants needed to make a forest. Without the pioneering role played by the birch, the ecosystems of the north would have evolved differently. The downy birch dictates the terms of what can grow, survive and move in the areas in which it takes hold. And that range is expanding fast as the Arctic heats up. Apart from humans, in the warming ecosystems of Europe's far north it is the downy birch that is setting the agenda.

It is hard to find my way in this gloomy world where everything appears a different shade of blue, but I finally locate city hall, a modern timber-clad building radiating orange light. The entrance vestibule is a two-stage affair, like a submarine airlock, where you must pass through a bath of blasting hot air. The receptionist is in a good mood. She, like everyone in Alta, is relieved. Finally there is some snow and finally the temperature is below freezing, even if only just. This is how winter is supposed to be.

'It gets very dark when we don't have any snow,' says Hallgeir, ensconced in his modern office lined with maps and stylish shelves. 'When I was young, my parents always said we had to be ready for winter by 10 October.' Winters have been getting gradually warmer in recent years, but the warmth in November and December 2018 is, he says, 'extreme'. The whole community has been in a state of panic, reindeer herders posting photos of a snow-less tundra on Facebook.

Hallgeir is a city dweller, a mild man with rimless glasses and a reserved air. He is also half-Sámi, the indigenous people of Arctic Europe who share DNA and a common linguistic heritage with the peoples of the circumpolar region from Finland to Russia across the Bering Strait to Alaska, Labrador and back to Greenland. The Sámi used to migrate across the land without hindrance, but now the 80,000 who remain find themselves instead citizens of one of four different modern nations: Norway, Sweden, Finland or Russia. They are the only indigenous group in Europe recognised by the United Nations.

Ever since the reindeer god first spilled the blood of the reindeer to form the rivers, seeded the ground with its fur to make the grass and trees, and cast the animal's eyes into the night sky for stars 10,000 years ago, the Sámi have lived in what other Europeans used to call Lapland, and what they call Sápmi (the land of the Sámi). Their rock art depicts a consistent way of life over millennia. Pictures carbon-dated to 8000 years ago show stick figures in boats fishing and the same figures hunting bear and elk and herding reindeer. Other pictures from the same location dated to 2000 years ago show stick figures fishing in boats, hunting bear and elk and herding reindeer. The only substantial difference is that the artist of 8000 years ago captured the likeness of the animals better.

Reindeer are central to Hallgeir's identity as they are for all Sámi. His mother's family were reindeer herders, but when his grandmother died in childbirth on the plateau, his grandfather brought his infant mother to the town, to Alta, and left her with a Norwegian family to raise. The grandfather went back to his herds beneath the wide skies of the plateau, to his *laavo* – a traditional tent much like a tepee – and married again. Hallgeir has a foot in both the city and the *laavo*. When I see him later that week at a Sámi cultural event, he is wearing the traditional Sámi felt jacket embroidered with gold, a silk scarf, reindeer-skin trousers and boots and an elaborately worked silver belt. He is an agent of the rational state, purveyor of bureaucracy and concrete, but he also has the winged blood of the nomad and longs to be governed not by man but only by the needs of the herd.

Reindeer are characterful animals, with their wide brown eyes, infinitely varied furry antlers, soft directional fur and enormous snow-proof padded hooves that clip clop in an awkward endearing gait. Their steady watchful gaze is uncanny and wise, suspicious and judgemental at the same time. Every reindeer has a Sámi name, and herders recognise every member of their herd individually, even by touch. Love is an insufficient word for the relationship. Codependency comes closer. The animals allow the Sámi to survive in an unforgiving world of cold and ice that would be death to anyone without clothes and shoes made of their skin. The people move because the reindeer move in search of grazing. Their whole culture has evolved around the migratory needs of the herds.

Birch is the herder's handmaiden. From shelter to fuel to transport, birch is essential to survival here. It provides tent poles. It was used for skis and sleds to allow the people to move from the lush summer pastures by the sea to the tundra on the plateau in winter. But the breakdown in weather is upsetting this cycle. The Sámi are among the first victims of climate change, forced to contemplate a little earlier than the rest of us the collapse of a whole culture.

The reindeer are the only pillar left of what was once a more diversified civilisation including forest Sámi who lived among the trees and fishing Sámi who resided on the coast. The forest Sámi lived in turf houses and disdained the profligacy of the Norwegians building their homes from timber – wood was for tools, boats and fuel only. But they are long gone, forced by the Norwegian government over a century ago to choose between reindeer husbandry or assimilation. Raising animals for meat was something the government appreciated, but subsistence in the forest served no economic purpose they could understand. The integration of the fishing Sámi has taken longer, but the collapse in cod stocks has helped accelerate the move to the towns, a process that it is Hallgeir's job to manage. Alta is a boom town of 50,000 inhabitants, growing as the countryside all around is drained of people.

Reindeer herding is valued by the rest of Norway and so it has persisted. The Sámi have always sold their meat to the southerners, and reindeer meat is an expensive delicacy that long ago became part of broader Norwegian culture. The Norwegian state sees reindeer as a farmed resource, with quotas and subsidies and strict controls on culls. To the official mind they are a commodity, a useful export from the otherwise unproductive vast plateau of the north, but for the Sámi the reindeer's significance is not only economic and cultural, it is also – as Hallgeir's leather trousers testify – symbolic.

'Reindeer are life. They are everything. Without reindeer, we die.'

And now reindeer herding, a way of life that has survived intact for 10,000 years, is under threat. This time it is not the Norwegian government that poses the greatest danger – although it plays a part – but the climate. Warmer winters are deadly for the reindeer in two ways: one is short and sharp leading to a quick death – ice; the other is slow but sure – too many trees.

Once upon a time the first snows of winter would fall some time in October, initially on the tundra, the plateau above the treeline, and then on the pine and birch forests of the river valleys and the coasts. Shortly after, the mercury in the thermometer would descend below freezing and stay there until April or May, when the snow would begin to melt and the rivers would rush with the clear turquoise of super-oxygenated ice. Until 2005 the average winter temperature was minus fifteen Celsius and it would reliably sink below minus forty at least once during the winter, eliminating even the hardiest of all insect larvae, a process that kept the Arctic pristine and pest free in the summer. This world of winter was dark and cold and dry. At those temperatures there was no moisture at all. The snowpack was the consistency of sand, made up of several layers of large snow crystals called *seaŋáš*. At minus forty or fifty degrees Celsius in the middle of winter, the quality and nature of snow crystals is critical to the survival of humans and animals alike.

Seaŋáš are crucial for a healthy snowpack with a balanced density – *guohtun*. *Seaŋáš* allow the reindeer to brush the snow away with their antlers, hooves and muzzles to reach the *Cladonia stellaris*, the carbohydrate- and sugar-rich lichen that grows on the ground in symbiosis with the grass of the tundra, a high-energy food for fast movement in winter. The lack of trees on the tundra meant that the fierce prevailing winds of Finnmark had a clear path across the plateau and blew the fine powdered snow into a shallow protective carpet that preserved the lichen beneath. Heavy drifts can crush vegetation.

But when the temperature climbs back up towards zero or, even worse, above it, this delicate winter ecosystem collapses. Even a little warming of the snow can create havoc. Moisture starts to appear in the snowpack at minus five or six, at which point it loses its sand-like quality, *seaŋáš* melt, and the snow starts to compact under the reindeer's hooves, ruining the grazing beneath. If the thermometer goes all the way into the positive, as it has done increasingly in recent years, it is a catastrophe. Melting snow or rain will freeze when the temperature goes negative again, forming a crust of ice over the ground, locking the vegetation away from the browsing reindeer. This happened in 2013 and again in 2017. Tens of thousands of reindeer died, some herders lost more than a third of their animals. In the last 130 years the thermometer has crept above zero three times during winter, twice in the last decade. From now onwards the projections say every winter will experience days above zero, meaning locking will become almost inevitable. Herds can be up to twenty or thirty thousand strong, and they are spread out across thousands of square miles of the Finnmark plateau, an area the size of Switzerland. Artificial feeding is impracticable, not to mention far too expensive. Something is going to have to give.

Reindeer are very intelligent animals. They are rightly suspicious of humans, wind farms, aeroplanes and vehicles. The increased human encroachment that is under way in Finnmark is shrinking their range further. They have the suspicion of prey. Their doe eyes, even bigger than their deer cousins', are always watching. Even while they are browsing on grass or lichen with a stationary head, a reindeer's eyes maintain a 300-degree view, missing almost nothing. A small tilt reveals the full 360. And most of that wide-angle view is in focus, unlike the pinpoint vision of humans or other predators, which has evolved to gauge depth and distance. Any sign of danger and reindeer will simply run.

They have an excellent memory for terrain and a perfect inner compass that tells them where to migrate in summer and in winter. They don't always know how to navigate the ravines and the rivers, but they have a homing instinct for the historical pastures they have known. And they have an inner thermometer too that tells them when it is time to move. If it is not cold enough to leave the autumn grazing and move to winter pastures, the herd will not go but will risk overgrazing a single area or range beyond its normal territory. And if there is too much moisture, meaning the ground locks, or the food supply is bad, a female reindeer can even con-sciously abort her unborn young, a trait shared with a few other mammals such as mice, monkeys and killer whales.

Warmer winters mean that the reindeer herds need more space in which to feed. Competition for the grassy tundra of the plateau is increasing from other reindeer, from wind farms, pylons, roads and mines. But the most formidable challenger looks much more innocuous even though, in the end, it will be the most effective: the humble downy birch.

The office next to Hallgeir's belongs to Tor Håvard Sund, manager of the Finnmark Forest Service. Tor is a large man in a checked shirt with an open face and a warm smile. He started out as a forest school teacher thirty years ago then followed his love of trees to become a tree expert and forester. We begin talking and immediately consult the huge map that forms one wall of his office, but he quickly gets frustrated.

'When was this map printed?' We locate the date in small print at the edge: 1994.

'This is totally useless,' he says. 'We need new maps. The treeline is out of control.'

Several interlinked factors affect the habitable range of tree species: the availability of sunlight, water and nutrients are prerequisites, but these interact with other variables such as wind and temperature in a delicate balance. Tiny gradations in altitude or latitude can mark large differences in vegetation, and these of course move. The tropics and the poles, with their more finely tuned weather patterns and ecosystems, are far more sensitive to global heating than the temperate zones accustomed to variable climates. The downy birch detected the current warming trend much earlier than most scientists. The trees were the canaries but few understood what they were saying.

The key change has been warmer and shorter winters. The Sámi have been saying for at least fifteen years that winters are getting 'weird'. The amount of light hasn't changed and the soil is the same, but more rain and more heat have made all the difference. The downy birch loves the warmer weather. It used to be confined to the dips and gullies on the plateau, out of the icy winds, but, unleashed by the warmth, it is storming over the top and out into the open, moving upslope at the rate of forty metres a year. An enormous amount of territory is being transformed from tundra into woodland at a lightning pace.

On the face of it, more trees might sound like a good thing. Except that the greening of the tundra is closely linked to more warming as the birch improves the soil and warms it further with microbial activity, melting the permafrost and releasing methane – a greenhouse gas eighty-five times more powerful than carbon dioxide in its warming effects over a shorter timeframe.

The downy birch is the rock star of the arboreal. It lives fast and dies young. It expends tremendous energy surviving in marginal environments but it cannot survive much beyond sixty years. The slower-growing conifers like the pine will take longer to establish themselves but they need birch to blaze a trail. Birch is a pioneer tree. In spring it can sense when the nights are getting shorter and the temperature is consistently warmer, and when it judges that the timing is right it flowers with two sets of catkins. The male is yellow and brown, like a furry caterpillar, and hangs in groups of four from the tips of the shoots. The female is short, green and erect. The birch is monoecious, like pine, meaning it can fertilise itself and doesn't need other trees to pollinate, just the wind. The spectacular wind of the Finnmark plateau is ideal for pollination, and for spreading the birch's fertilised seeds, when they come, in autumn.

After pollination the downy buds covered in fine hair break open to release more than a million little winged seeds like hard beetles onto the wind. A good year for seed dispersal is called a mast year. Every year is a mast year these days. The little seed cases lie dormant under the snow until the spring light and warmth trigger germination and the seed puts out a root tip seeking pliant soil. More of them than ever are taking root among the matted moss and lichen of the tundra. Before, even if seeds successfully germinated – not a given by any means – they only had a brief window (from June to October) to come into bud and leaf and build up the necessary resources to survive the punishing deep freeze of the polar winter. Birch are especially good at this – amassing a tough woody bark in record time and expending considerable energy on producing oils and proteins to withstand the cold. Even so, the freeze usually killed off most of the saplings that had struggled up on the tundra and maintained the treeline at a certain altitude. Before, the growing season was May to October; now it is April to November, and it is wetter both in summer and in winter. Ideal conditions for birch.

The reindeer herders joke that these are also ideal conditions for the birch's parasites. The larvae of the birch-eating autumn moth, *materjokt* in Norwegian, die at minus thirty-six Celsius. In recent warmer winters, when it hasn't got that cold, they have survived to wreak havoc over thousands of hectares of birch forest, but not even these insect outbreaks have made much dent in the birch's assault on the tundra.

'Nature is complicated,' Tor says. He has already witnessed big changes in his own lifetime such that he believes nature is finding a new equilibrium, testing the permutations of natural selection, trying different options in search of balance.

'Sooner or later, the whole of the plateau will be covered in trees.'

There were trees here before of course. Before the last ice age the forest reached all the way to the coast. The trees are only part way through their journey back, but a journey that took millennia is now being travelled in decades. The speed is a problem. Most species cannot adapt that fast. There has always been a relationship between the tundra and the forest, and the Sámi have existed at the interface. But the word 'forest' meant something different before warming began to show its hand fifty years ago. Previously it was a slowly moving, evolving landscape with birch at the fringe creeping ever northward, the birch sheltering the young vulnerable pines. The old-growth pines created a diverse forest full of hundreds of different kinds of plants that the reindeer could eat. There is almost none of this old-growth forest left in Norway although large swathes persist across the border in Finland.

It takes 160 years for an old-growth pine and birch forest to form. One that is suitable for reindeer to graze in. Young pine trees shed too many needles and suffocate the ground-growing lichen. After thirty years, dense saplings and small trees create a humid microclimate in which lichen declines and mosses in the canopy flourish. The new awakening of lichen occurs in natural-state forests after a hundred years, when the forest has thinned out. The lichen that grows in these old-growth pine forests is like coral, blooming in great spongy fronds and hanging from the dead branches of old trees. Alongside its enhanced ability to absorb and store carbon, this lichen is the most important difference between young and old forests and why old forest is crucial as an alternative to the tundra for reindeer winter grazing.

Is this also why an old forest feels good to the human soul? When we walk among a rich inter-generational community of trees accompanied by a profusion of other species, there is abundance and richness, but there is also satisfaction at the sight of nature having achieved a natural balance. As one reindeer herder campaigning to protect Finland's old-growth forests from being logged told Greenpeace, 'You don't go to a young forest to feel revitalised, you long for an old one.'[1]

In Norway aggressive tree growth is now creating havoc. The birch is racing over the tundra faster than the pines can keep up.

This is bad news for the reindeer and the humans that rely on them. Upright birch forests don't develop a canopy, they are more like thickets. Without a canopy, they trap more snow, their branches bending and cracking with the weight and their mass forming a windbreak for drifts too deep for the reindeer to walk or dig through. Their roots warm the ground below, causing ice and melt around them. In time a hectare of birch will deposit three to four tonnes of leaf litter on the ground, further improving the organic composition of the soil and encouraging other plants. Reindeer do nibble the twigs of young birch, 'but even if you doubled the number of reindeer in Finnmark county you could not stop the birch'.

Tor smiles ruefully. He is a scientist; emotion has little place in his office. He accepts the advancing treeline as inevitable. For him there is an upside. Pine is what Tor really cares about; birch, essential to the Sámi, is not nearly so valuable in a modern economy. The forestry industry of Finnmark is based on Scots pine, and is only just coming back to life after a series of catastrophes. The life cycle of pine trees is long. Ten generations of birch can live and die within the lifespan of a single pine. Like their cousins in Scotland, some of the pines that are alive today have witnessed swathes of human history from the Viking voyages to Greenland in the Middle Ages onwards. And for them the twentieth century was the most traumatic time.

In the 1930s the largest sawmill in Europe was in Kirkenes, a Norwegian port on the Barents Sea, not far from Murmansk across the Russian border. Kirkenes consumed 130,000 tons of pine a year. When the Nazis invaded in World War II, however, they burned it down. Instead they pillaged Norway's forests for shipbuilding and export on an unprecedented scale. And when they retreated, they razed all the wooden towns of Finnmark. Tor's mother remembered watching from the hill as her home town of Vardsø burned to the ground. The people of Finnmark still refer to the destruction as a kind of holocaust. After the war, what remained of the ancient forest was felled to rebuild their homes.

The pines of Finnmark therefore are young, Tor explains, mostly not more than sixty years old. They are not ready for harvesting until they are 120. After that they grow more slowly. They can live to 300 or 400 years old, but there is, as Tor puts it, not much 'value added' after 120. The forests of Finnmark will never grow old. It might be possible for the reindeer to survive in a changed landscape, but only if the forest is given enough time to mature and develop, time that it is not clear the forest has. In any case it is so long since they have grazed in an old forest it seems even the Sámi don't know what they have lost.

'The Sámi in Norway have never known old pine forests like the ones in Finland; they have grown up with forestry,' says Tor.

Tor is supposed to manage his forests in consultation with the Sámi reindeer herders, and every year more and more herders beg him to cut the pioneering birch to protect the precious tundra habitat needed for reindeer. The herders who traditionally considered themselves a part of the natural world, not distinct from it, are fighting a losing battle against nature. Tor is blunt.

'The Sámi will need to find another lifestyle.'

In spring and summer the Sámi bring their herds of reindeer to the coast, the majestic jagged hills that serrate the fjords, or to the islands off the coast that stud the Barents Sea like rough-hewn jewels. It used to be common in springtime to see herds swimming across a fjord to reach the lush grass of an untouched island, the herders and their dogs following in kayaks or rowing boats. These days most herds make the crossing in ferries otherwise used for cars.

In summer many Sámi are dispersed with the herds, living in *laavo*, their traditional tents made of woven wool stretched over an interlocking pyramid of birch poles. Children, released from school for the holidays, will still often spend weeks at their family's summer place, rarely venturing home. It was only recently that herding families began to settle predominantly in one location, required by government edicts to live by a road and to send their children to government schools – an attempt to clip the wings of the nomads and keep them where they could be seen, and their animals taxed. Before, herding was a family affair; now it is mostly a male activity as women look after school-age children.

In autumn and winter, though, the herds return to the plateau, to their 'winter place' – an area that a family grouping (called a *siida*) has been coming to since anyone can remember. It is during winter that Sámi socialising takes place, when herds are gathered on the plateau mostly within striking distance (a day's hard riding by snowmobile) of the centre of Sámi cultural life, the town of Kautokeino. Its name in Sámi – Guovdageainnu – means 'the middle', and it is literally the middle of the Finnmark plateau, or – as some joke, since the plateau is characterised by its emptiness, devoid until recent decades of trees and any permanent human settlement – the middle of nowhere.

It is Kautokeino that hosts Sámi University College, the Sámi Cultural Centre, the Beaivváš Sámi Theáter and the International Centre for Reindeer Husbandry. For the hub of Europe's oldest continuous civilisation – a way of life essentially intact for over 10,000 years – it is surprisingly small. There are only 1500 permanent inhabitants. Photos from the 1950s show the buildings of Kautokeino surrounded by the unbroken white of snowy tundra without a tree in sight; now it is in the middle of a birch forest.

From Alta, I take the road to Kautokeino, 250 kilometres directly south. It starts among the mixed pine and birch forests that border the wide pebbly curves of the River Alta, the world's finest salmon river, so they say. Then it climbs swiftly through a narrow gorge beneath sheer towering cliffs hundreds of metres high, up onto the plateau above. A clear stream roars into clefts in the rock beside the road.

The Sámi traditionally worshipped rocks, trees, rivers, mountains and other sacred sites, sacrificing fish, white albino reindeer and other animals to ensure good hunting, fishing or simply luck. They talked to their surroundings; the animals and plants were their community. Oneness is central to Sámi cosmology. There is no concept of 'man' or 'nature' at all, only a circular system

depicted on the Sámi shaman's magic drum: in the middle a rhom-
boid sun with four rays, the ruling gods of thunder, wind and the
moon, the 'dog of the gods' – the bear – and on the lowest tier
the Sámi, his home, his reindeer and the birds and game of the
forest. The divine forces that manifested themselves through the
gods demanded respect, and when Sámi passed sacred sites they
would dress in finery, perform a song or bare their heads as a mark
of veneration. There are people still alive who remember being
told by their parents to say good morning to a particular tree or
rock.

The gorge carries a spiritual weight. It has the echo of a
cathedral – or a forest – and must surely be a sacred place. It is a
reminder that names, stories and spirits once covered all the territory inhabited by humans. For most of us, those voices are distant
echoes lost to a mysterious past, but for the Sámi they are only just
out of earshot. The land has not gone quiet. It's just that we stopped
listening.

At the head of the gorge the water frets over pebbles and then
lies still, widening and widening with the view as the valley expands
like a door to an upper world swinging open. The snow here is
thicker, deeper than at the coast, but the river is not yet frozen.
Then, after fifteen minutes or so of twisting and turning along the
river's edge, a faint line of ice bisects the slow expanse of water at
an angle, like a pane of glass. Water slides over and under it. This
is the long-awaited, much delayed, beginning of the freeze. From
here on, the surface of the river is opaque and hard. Swirls of white
and shards of blue flash in the sharp morning air. To the south is
the pink and orange furnace of the dawn. Without the mountains
to hide behind, the sun is closer, although still coy, keeping just out
of sight. Shots of red and yellow streak the sky above like tracer
fire, and the speckled hillsides, snow and shrub, hum with a warm
rose glow.

Ever since Alta, one hundred kilometres behind, all along the
roadside shrubby birch have kept close company with the car as
they will all the way to Kautokeino. Only once, when a pronounced
mountain rises above the level of the open river valley, is there a

flashing glimpse of unforested tundra: smooth unblemished snow cut by a line of bent and twisted little figures, a battalion of birch marching grimly upward. The trees here are not as tall as at the coast, rarely reaching above two to three metres high. Their silver bark is more disfigured too: not the smooth papery coating of straight vigorous trunks but a gnarled skin pocked with blisters to protect the trees from the cold. The white colour comes from periderm, a powdered surface layer which reflects sunlight and protects the tree from sunburn as well as from the low winter sun, which could unfreeze the pressurised sap in the trunk and cause the tree to burst. It is also used by the Sámi as a medicine. This is krummholz – stunted, slow-growing trees surviving at the outer edge of their natural range. Some question whether they deserve the name tree at all. But is a hundred-year-old bonsai, half a metre high, not a tree? Or is it even more deserving of respect for its incredible feat of survival?

A short distance from Kautokeino the road crests a ridge, and below the plateau unfurls in a wide vista of black – the trees – and orange – the sky reflected on the snow. Threading its way through the centre of the scene is the line of a snaking river, here and there flowing again, unfrozen, its surface shimmering like liquid gold. The eerie almost-dawn has passed into the moment just after sunset without the sun making an appearance at all. Half the sky is on fire. It is one o'clock in the afternoon. We are at the threshold of a twenty-hour night. From this vantage point the plague of trees is frighteningly clear. As far as the eye can see the tundra of the plateau is flecked with black streaks like the speckled breast of a snowy owl. The pattern has been painted by the prevailing wind, carrying the hard seed cases on their tiny wings in gusts and currents over the rolling hills. Out of the wind, where it has dropped seeds in the folds and the hollows, the birch is taller and thicker.

It is a beautiful scene, but the fact that the trees shouldn't be there and the river should be rock hard with ice several metres thick by this point in midwinter, capable of sustaining the weight of a herd of reindeer or an articulated truck, makes the beauty of the vision hard to absorb. Perhaps it would be different if we didn't know how it used to look, if we could pretend it was a freak year and not part of an accelerating pattern. In fact, this stretch of otherwise unspoiled territory is in the midst of massive upheaval. On this winter day, at this spot in the Arctic Circle, at minus one degrees (fourteen above average for this time of year), it is hard to avoid the feeling that if there is a tipping point in earth's climatic equilibrium, we have already left it far behind.

In a yellow one-storey house on the outskirts of Kautokeino, Berit Utsi holds her two-year-old son to her chest and looks out into the mounting dark at the lake covered by a paper-thin sheet of ice and ringed with birch trees. She is the secretary of the local reindeer herders' association and she has agreed to talk to me about the problems caused by the advancing trees. She has a calm expression, but there is a restlessness that flickers now and then behind her eyes; her anxiety cannot be completely hidden.

'It's not our culture to make a drama,' she says. That is an understatement. The famously reserved Norwegians have nothing on the Sámi, for whom emotion registers only as a tremor or a quiet laugh, a faint line in an otherwise frozen expression.

'Everyone kept a calm exterior but inside we were all very worried.' She is speaking of the incredibly warm winter, which has just been blessed with its first snow. But Berit's worries are not over. Her husband is still out there, somewhere. She doesn't know exactly where. He moves about a lot, and the mobile phone signal is often poor. This is a very stressful time for reindeer herders even in a good year: moving the herds from autumn to winter grazing, keeping the herd together over hundreds of square kilometres.

The change in colour of the tundra has serious consequences. Reindeer are the only mammals that can see ultraviolet light – invisible to humans. In the low light of the polar night, when the sun doesn't rise, this ability is crucial to their survival. Lichen absorbs UV and so appears black against the snow. There is also emerging evidence that lichen fluoresce in different colours and so might be visible to reindeer through the snow.[2] Their eye has a special strip called a *tapetum lucidum*, common to nocturnal animals and insects, a 'bright tapestry' that absorbs light and reflects it back into the retina to improve their vision in low light. Unique to reindeer, in summer the *tapetum* is golden and in winter it turns deep blue to absorb the UV. In unbroken snow the reindeer are calm and usually stay in one place, digging the snow to reach their food. But a speckled black and white is tempting and confusing, raising the possibility that more easily accessible food might be on offer. They avoid digging and instead nibble the grass and lichen uncovered at the base of trees, moving over much greater distances and causing a headache for the herder, who must keep an eye on them and keep them together, to avoid straying onto the territory of a neighbouring *siida* or, worse, mixing with another. Separating 10,000 reindeer can take two weeks.

Apart from last week, when he came back for a few days because she had had an operation, Berit's husband has been out on the plateau with the herds for two months straight. The family's entire income and savings are invested in the herd. One animal is worth over 1200 euros (£1100) at the slaughterhouse, and every part of the carcass – skin, antlers, hooves and sinews – is used by the Sámi for clothes, tools and handicrafts. The high stakes encourage risk.

'There have been a lot of accidents lately,' Berit says. A 'point check', driving a perimeter all around the herd, is the daily routine of a herder. In cold conditions the tracks of an errant animal are easily visible on the snow, and a snowmobile can fly over open tundra, frozen lakes and rivers in a thirty-kilometre circuit. An unfrozen landscape pocked with shrubs is much harder to negotiate. Without sufficient snow for a snowmobile, and without ice, a herder on a quad bike must go around lakes, rivers and trees, sometimes up to sixty or seventy miles further. It takes a whole day, burning a lot of fuel and crushing a lot of lichen that takes hundreds of years to return. And the next day it must be done again.

'People have been driving snowmobiles on stones, hitting trees and crashing, ending up in hospital . . . or maybe the ice is strong enough to carry the reindeer, but the quad bike falls in. Sometimes you take a risk – it is too far to go around. Last year two people went through the ice and did not come up,' says Berit.

When she was a teenager, Berit tried working in a town but it didn't feel good. She missed her reindeer. She grew up with them, spending every summer with her family and the animals in the Lyngen Alps, near Tromsø. She remembers the tundra with fewer trees when she was a child. She feels the change as a loss, but like most Sámi I meet she is pragmatic: 'We adapt, we always have.' But the changing weather and the advance of the trees combined with other pressures on grazing – roads, mines, wind turbines – mean that the economics of reindeer herding are becoming harder and harder. And, to make matters worse, the government is aware of the shrinking grazing and demands ever larger culls of animals every year. Her family needs another income.

Berit would love her children to have the chance to become reindeer herders should they wish – it is a strong tradition – but if the children do not live with the herds full time, as her parents did, then the knowledge they accumulate will necessarily be less. And it is not just the ways of the reindeer that must be learned. Nights in the wild were also when stories were told and tools made.

The *soahki*, as the birch is called in the Sámi language, is almost as essential to traditional Sámi life on the tundra as the reindeer. It was crucial for shelter: for making the poles of the *laavo*, and for insulation – fragrant birch twigs were laid on the floor. The wood was essential for transport, used for making sleds, skis and snowshoes, and for fuel. And in autumn, like all trees, the birch decreases the moisture content in its xylem – the inner part of the trunk – ready for hibernation. This means that birch harvested during winter will burn well even if unseasoned. Its tannins and oils are used in treating clothes and skins and making oiled paper. Its bark was used for canoe skins and fermented in seawater. Tannin was also used for seasoning the wool, hemp or linen sails of traditional boats. In spring the sap can be tapped to give a mineral-rich drink or else to ferment as the basis for a kind of mead.

'And in autumn, mushrooms!' Over seventy kinds of fungi coexist within the habitat of birch roots.[3] The fate of the Sámi and the birch have ever been linked.

Appearing as it does as the harbinger of spring, the birch has always had a role as a fertility symbol for the Sámi – and elsewhere too. In Scottish folklore a barren cow herded with a birch stick will become fertile, and further south in England a maypole was traditionally made of birch. A 'besom birch wedding', when a couple jumped over a bundle of twigs to be married, was a common alternative to English church weddings until recently. It was also associated with purifying and cleansing, hence the 'beating of the bounds' of a parish was always done with birch twigs.

'The birch is our friend!' Berit says as if feeling bad for maligning the tree upon which she relies so much.

Out there on the tundra, the trees are making her husband's life difficult and dangerous. Here, in the kitchen, birch is everywhere. Berit's modern kitchen is still full of the traditional handicrafts of the nomads, made on her summer trips to the mountains. Her wooden spoons and ladles are all carved from birch – 'much stronger than pine'. Cups and bowls on a shelf are also carved from birch, while the handles of handmade knives are of antler and bone. A coffee bag made of worked reindeer skin hangs beside the kettle, and on the side sits a hat made of fox fur and reindeer skin. Her son wears booties made of reindeer skin insulated with hollow-stemmed tundra grass. And in a small pot on the worktop are shavings of birch bark for tisanes and medicinal brews.

'But now the trees have become too much,' says Berit, frowning. She is studying to become a teacher.

As in Scotland, the balance of grazing animals and trees has been upset and the humans are confused. But, as with Scotland too, the die has been cast. Emissions already in the atmosphere will determine the future shape of the forest. The task now facing us is the same as that facing Berit: the struggle to accept what is happening and adapt.

Kautokeino seems to be a town in hibernation. Or else not really a town at all, just an elaborate fabrication like a film set. At the Arctic Motel a knock on the door in the darkness is in vain. At the youth hostel a lone elderly woman peers out of the only lit window in a concrete 1970s edifice and waves for me to go away. The door of the craft centre is open but it is deserted. Upstairs, three elderly people in very large coats are reading newspapers. They point back downstairs to the workshop. In the workshop a man with ear defenders clamped to his head is lost in concentration at a lathe; even if he can see me he is not about to stop his work. Shops appear to open according to their own caprice, and, like Alta, there are no pedestrians on the streets, only empty cars outside houses, their engines running to keep them warm, or on the roads mysterious slow-moving cars with brake lights blinking in the dark.

The following day will be the fiftieth anniversary of the founding of the Norwegian Sámi Association, a political group that presses for Sámi rights. Perhaps everyone is lost in preparation for this important event?

'What?! No!' The two women drinking coffee in their kitchen laugh and splutter. 'Nobody cares about that!'

Mārijā is wearing a black watch that matches her black nails and black and white chiffon blouse. Her necklace and earrings are gold, and her partly shaved hair is dyed red. Sara-Irene also wears a black and white shirt, and her hair is also shaved at the sides although the fringe is dyed blonde. She has three pearl earrings in one ear. On her middle finger Sara-Irene wears a ring made for her by her husband; it is the 'ear mark' of her reindeer herd. The animals of every herd have their ears cut in a particular pattern to denote ownership. In the corral, at counting time, an experienced herder will know her animals in an instant just by feeling the ears. Sara-Irene is immensely proud of her reindeer. She won't reveal how many she has: 'It's not polite to ask.'

'She has loads,' Mārijā says, laughing.

'But not as many as Mārijā,' Sara-Irene replies, and they both collapse into giggles.

They are not into politics; they have the indigenous people's traditional disdain for government: 'The government doesn't care about the Sámi.' There is no point in trying to influence things by going to the meeting. But Mārijā is the head of the welfare office in Kautokeino, so she actually works for the state?

'Yes! Imagine! The enemy!' And she laughs herself into a cough-
ing fit once more. Her great-great-grandfather was beheaded by
the Norwegians in 1852 for standing up to the government. She
has inherited his ear mark. It is a noble ear mark, a famous one.

When I bring up climate change, the mood turns. 'The government just wants to control us and our herds. It's just an excuse . . .
forcing us to cull our animals. Climate change is bullshit. When my
mother was smaller the weather was exactly the same as it is now,
ninety years ago. So I'm not worried. We've seen this before.'

She cannot explain the trees. When I press the point, she mutters
curses about the Norwegian government and makes Sara-Irene
laugh with a joke that I don't understand before nipping outside to
the patio for a cigarette. When she comes back in, the conversation
is firmly moved along.

Reindeer herders are not separate from the herd but a part of it.
They can sense, days or weeks in advance, when the herd will
move. They cannot eat unless their herd has eaten. Contemplating
threats to the welfare of the herd like climate change must be akin
to imagining your own children starving or dead. Denial, of course,
is the first stage of grief.

Sara-Irene and Mārijā both work in the town. Sara-Irene runs a
nail bar. She is the one responsible for Mārijā's black nails. But
there are times of year when they will not miss being with the reindeer: the time of calving in the summer and of ear-marking young
calves in the late winter, when the whole family is needed to corral
thousands of reindeer day and night.

'Sometimes,' says Sara-Irene, 'we feel schizophrenic!' Then both
women laugh like crazy. They may be good at their jobs, take pride
in their work even, but their soul is not in the town. It is out there,
in the hills, running free with the reindeer.

The next morning the town is half-asleep again, deadened by the
dark and the cold. It is now minus eight, still not cold enough, the
woman in the guest house complains. The sky is overcast, and
without its clear dome, the light is a kind of murky soup. The river
beneath the bridge is still partly liquid, moving in a slow sweep

past the dark church on its spit of land. Lights gleam in the houses; there are still hardly any cars and no pedestrians.

But the petrol station is different. The forecourt is blazing with white lights. Queues of huge pickup trucks, many outfitted by the same 'Arctic Truck Co.' with enormous snow tyres, sit with their engines running, filling the crisp air with clouds of diesel fumes. Behind each one is a trailer carrying a snowmobile or quad bike or both. I drink a coffee perched on a steel stool like at an American diner and watch as men wrapped head to toe in snow suits and fox-fur hats clamber down and fill batches of jerrycans with fuel. They move quickly, with purpose and energy. They enter the shop with long strides, slap money on the counter, greet the staff loudly, fill up their arms with snacks and sugary drinks and shout goodbye. Then they jump into their massive polluting machines, push them into gear and roar off into the murk that passes for morning. They are the reindeer herders, off to do their point checks. Some might be back tonight; some might be gone for weeks; some might not come back at all.

There are images of reindeer everywhere in Kautokeino – on the walls of shops, in the logos of government agencies and Sámi cultural organisations. They are in the craft shop, on the postcards, depicted in the lights outside the supermarket. But there are no actual reindeer. In three days I have still not seen a single one.

If Kautokeino feels soulless and weird, it is because the town is not the centre of life on the plateau. Kautokeino's historical function has always been that of filling station, a stopping-off point with essential services. Apart from the fast-food counter in the petrol station there are no restaurants, no shops aside from the supermarket and just one craft boutique for the trickle of tourists who make it up here, mostly in the summer. People's identity, their wealth, their husbands, the well-being and the future of their family and their culture, lies with the fate of the herds out there on the tundra.

Kautokeino might seem like a sleepy backwoods town but upon closer examination it seems to be in the grip of a kind of collective neurosis. Everyone is aware of the pitched battle between man and nature taking place in the hills and tundra all around but almost no one is willing to acknowledge the truth: this can't go on in its current form. It is a microcosm of the struggle playing out across the world. On one plane exists the prosaic patina of daily life in a hydrocarbon society: going to work, to school, to the supermarket, filling up with gasoline, oranges and mangoes from far away, and yet, right below, on another plane exist the spirits of the tundra and the forest shrieking their warnings.

The Sámi Kingdom of the Dead, Jábmiidáibmu, was located underground and ruled over by Jábmiidáhkká, the mother of the dead. A river of blood separated the land of mortals and the mythical reindeer land but was ever present. As with ancient Celtic culture or contemporary indigenous peoples around the world, the role of the *noadi* – shaman – was to enter into dialogue with the ancestors and the spirits and seek their advice. But the *noadi* are long gone.

In the only hotel in town the youth wing of the Norwegian Sámi Association is having a precursor event before its main annual meeting. Young activists in beautiful traditional costumes and matching trainers sit cross-legged on reindeer skins, behind faux police tape that reads DECOLONISED AREA, their iPhones at the ready. The title of the meeting is 'The Future Is Indigenous'. It is a tempting and romantic notion, but for the most part the earnest kids sitting on the floor are not from Kautokeino and not from reindeer herding families, nor do they have any wish to live in *laavo* and follow reindeer on foot through the snow for half the year.

Everyone knows someone who has given up their reindeer. Márijā and Sara-Irene swear that they never will and pity those who have. Who would willingly give up this sacred and ancient way of life, this birthright? Those who continue are either the herding aristocracy like Márijā, who are so rich in animals that they can weather the storms for the moment as well as purchase second homes in Tromsø and third homes in Oslo, or else they are true devotees: possibly addicts, possibly mad. I am not sure which epithet best describes Issát, but his experience perfectly captures the cognitive dissonance forced upon us by warming. Rationally, we know what is happening and what is likely to happen. But practically and emotionally it seems we will do everything we possibly can to avoid accepting the facts.

I meet Issát in his nondescript office in the back of a municipal building in Kautokeino at 9 p.m. at the end of a long day. His organisation, Protect Sápmi, is an NGO that provides legal advice to Sámi communities challenging the takeover of their land by multinationals and government parastatal organisations, and it is overwhelmed. The warming Arctic has led to massive interest in 'opening up' the north not just in Norway but all over the circumpolar world: Russia, Greenland, Alaska, Canada. Norway is self-sufficient in renewable energy but there is massive demand from Germany, the UK and the Netherlands, and wind farms in the Arctic Circle are rapidly colonising the few remaining tree-less mountain ranges in Finnmark. The Sámi people are supposed to control 96 per cent of the land of Finnmark according to a recent law, and the Norwegian government is supposed to follow the UN principles of 'free, prior and informed consent' for the alienation of indigenous land, but it doesn't. The only organisation that does abide by international law is, strangely, NATO, which has leased large tracts of Sámi land for military manoeuvres.

At the end of our discussion at around 11 p.m., when I am ready for bed, Issát announces that he will now begin his 'second job', reindeer herding. He invites me to come along. His home is up the hill, a terraced house in a small housing estate resembling many others in Europe. While I wait outside Issát goes inside to kiss his wife and his four sleeping children and then put on his reindeer herding clothes: two pairs of thick wool socks, thermals, down trousers, a fleece, a knee-length outer coat, a snowmobile jacket bearing the logo SINSALO, thick rubber snow boots, mittens and a battered old reindeer-skin hat lined with fox fur. He emerges ten minutes later. Without his glasses and suit and neatly cut hair he is transformed. No longer the quiet, diffident legal expert; he has become an action man.

Outside, it is only minus five, but we have to be prepared to be out all night if an animal is lost or we have an accident. Recently, a herder was trapped under a snowmobile for twelve hours before his friends came looking for him. Issát whistles to his dog, who jumps up on the back of his quad bike next to me – she knows where we are going.

Through the dark streets of terraced homes on the edge of Kautokeino, the quad takes us out of town, past the scraggy birch struggling up the hill until the clumps get shorter and shorter. We speed past the '60' sign riddled with bullet holes and up onto the plateau. At the top the trees are only head high. Issát slows down and steers the quad to one side of the road. Standing up, he peers into the beams of his headlights tracing the edge of the tarmac. The line is marked with red poles. Should the snow come as usual, these will delineate the edge of the road. He is looking for tracks.

Where the snow is disturbed he moves especially slowly. Marks in the snow mean his reindeer have crossed the road and strayed. The trees cause the reindeer to roam more widely, which means more conflicts over territory and grazing areas and more disputes with neighbours. Issát must patrol every night to make sure his reindeer are on the right side of the road. The changes are threatening to tear the Sámi community apart. They also put incredible pressure on the herders and their families.

'Does your wife not mind you being out every night?' I ask.

'She's used to it,' he says.

Issát is concentrating on the snow, on the reindeer that have trespassed. The question he is trying to answer is when? He points and shouts and talks fast and fluently, in English, not at all like the reserved man in the beige office who consulted charts and struggled for the right words only an hour ago. Out on the plateau under a sickle moon, astride his powerful machine, he is exuberant – a different person, blue eyes dancing in the headlights.

He kneels to examine the crust of snow on the tracks and their direction. There are seventeen Sámi terms for snow crust, and seven grades of hardness. Survival depends on precision. The tracks are old.

Back on the bike, we speed on across open ground – *oppas* – an area untrodden by reindeer, quartering the snow like a barn owl, looking for tracks. Issát spies one, then many, heading in the wrong direction. He swerves at speed, following the tracks. The quad bike briefly leaves the ground then lands with a crack on a frozen lake. In the headlights the tracks lead straight across. Issát holds his breath as the ice creaks and strains, issuing an occasional report like a gunshot. Twice in the last month he has gone through the ice. Last time he got soaked in a shallow pool up to his chest, and the bike had to be winched out, taking several days to dry out in the garage.

'This is the most dangerous job in Norway!' he grins. And it's true: herding is more dangerous than working on an oil rig or in the military.

The stars are out in a majestic show. Issát is pointing and shouting, giving the Sámi names for the stars, but I cannot hear a word. The wind is roaring in our ears, the bike is flying over tundra, bouncing over rocks and boulders and then through a birch forest, the twigs whipping our faces.

After an hour and a half, at nearly two in the morning, Issát slides the bike to a stop.

'They should be here.'

'Do you have GPS?' I ask.

There are ten reindeer in the herd tagged with GPS, but Issát's phone is out of juice. In any case, he prefers to do it this way. His instincts rarely let him down.

He turns the engine off and the lights, and listens for the bells that some of the reindeer wear. The silence is immense. Nothing. The stars blaze so intently we could touch them. The limbs of the trees imprint their shadows on the snow.

'Oh well,' he says, turning the key and twisting the bike towards home.

It seems unbelievable, all this effort only to give up at the last. Issát tells me his brother can continue the search in the morning. I am shocked, but I realise I have misunderstood the purpose of the excursion. Tearing through the night, alone in the silent expanse, tracking wild animals in a game of life and death – it's like some high-stakes computer game masquerading as a way of life. Finding the reindeer is not always the point. These are the actions of a man watching his inheritance dying in plain sight. Issát knows that herding reindeer this way is no longer viable. He spends all day arguing with the government and mining companies for compensation on the basis that it is, but at night he dreams the opposite. As the quad bike whines down the hill back towards the sleeping town glowing with neon in the valley below, the trees by the roadside gradually increase again in height and the howls of the dogs of Kautokeino fill the night air. A wolf has been sighted nearby in recent days – another consequence of the expanding forest. Issát pulls up outside his house shuttered in darkness, and I climb down, stiff with cold. As he unwraps his outer clothes and goes inside to bed, a light comes on next door in his sister's house. Her daughter, his niece Māret, is just waking up.

It is the day of the big meeting, the fiftieth anniversary of the Norwegian Sámi Association. Māret is a chef and she is cooking for the 200 delegates. She needs an early start.

She makes her way across town to a huge blue and black rectangular building next to the supermarket, the sports hall where the meeting will be held. To one side is a commercial kitchen fitted out in stainless steel. Māret dons her white chef's outfit but combines it with a *gotki*, the traditional multicoloured four-cornered hat of the Sámi. Māret is famous among her people. She is one of a few Sámi chefs trying to preserve its cuisine and traditional practices around food and the medicinal uses of plants.

'I want to make people think through their stomachs!' she says when I join her later after a brief moment of sleep to recover from the manic night. Her round face breaks into a wry smile and her eyes glitter beneath their heavy lids. She looks tired.

'I am very tired, and very angry, but I haven't given up. I can make a protest through my food. Everything is from nature.'

She is cooking reindeer. Her own reindeer slaughtered according to biodynamic principles, when the moon is growing. She can be sure of the taste. You can taste the changes in the meat, she says: the wetter climate, the encroaching trees and the decrease in lichen in the diet of the reindeer. Winter animals from the plateau used to be the most prized, with the leanest meat, the best combination of muscle and fat. But now most reindeer meat tastes of the coast – too much grass.

After reindeer soup will come pancakes made with birch flour and black pudding made from reindeer blood. The inner part of the birch bark, the red cambium tissue, is dried and ground into flour, which is sometimes used alone and sometimes mixed with spelt or wheat or other flour. It has a woody aromatic taste. There is also bread made with dried reindeer blood, reindeer brains and flour from ground pine bark. This is an important source of vitamin C and other minerals that meant the Sámi never suffered from scurvy despite shortages of vegetables during the long winters. This is one of the reasons the pine, called *bèahci* by the Sámi, is sacred – it is necessary for life.

'This indigenous knowledge is very lost. But it is the key to our survival in this polluted world,' says Māret.

Getting the blood was difficult. She is not allowed to serve it fresh, and the manager of the local slaughterhouse, although he is a cousin, suffers, she says, from 'assimilated thinking'. That is to say he is a prisoner of the Norwegian government's regulations.

The remainder of the menu is cod caught at Alta, meatballs made from elk and fried seaweed, an old Sámi delicacy.

Māret's husband's family are from the coast, where she lives now, but she grew up in Kautokeino. 'When my father taught me to drive a snowmobile, he told me to aim for the single birch tree on the tundra as a landmark. Ha! Now I would not recognise the place.' But she doesn't use a snowmobile any more. She is teaching her children to herd on foot, with dogs, in the old way. She does not want to burn any more petrol than she has to.

'As a human you are not allowed to destroy the food or habitat of other animals. You only take what you need because you are not alone.' This is the Sámi concept of sufficiency, called *birgeju-pmi*: you only take what is necessary from nature, never a surplus. It is the exact opposite of the modern idea of sustainability, which is based on the maximum surplus that can be extracted without destroying nature's capacity to sustain the resource. It is an important distinction. Understanding nature and using it well was not only essential to life in the Arctic but a value in itself. But *birgejupmi* is being lost, according to Māret, because of 'this money thinking'.

She prides herself on learning and preserving the old knowledge. How to tell if a birch tree is diseased and knowing that a sick birch tree will produce antibodies that are beneficial to sick humans too. To read the skies and the plants to tell when it is going to snow, when it is going to be dry and for how long the snow will lie.

'I read nature and then I check my apps on my phone and I see: I am right!'

Lightning in the autumn means the early summer next year should be warm. And the colour of the birch leaves in autumn and how long they take to fall hold clues to how long the snow will lie in spring. This year the colours were bad: they were green and brown with no red and yellow, not the classic, beautiful array of yellow and red. Māret fears it'll be a wet and long winter rather than a cold, dry, short one.

Her own forecasting is based on the evolutionary memory of the trees themselves. The reason birch leaves go so spectacularly red in autumn is because the tree is preparing for winter by withdrawing all the chlorophyll from its leaves, storing it for the spring, locking out excess moisture from its trunk and shutting down the capillaries in the phloem, the inner bark that acts as its blood vessels. The remaining carotene is what gives the red stain. But if a tree is too wet or conditions are too warm, it hedges its bets, leaving the leaves on its branches for longer, keeping some chlorophyll active, allowing them to gather as much energy as possible. It is as if the tree knows that the winter will be long.

'I am not angry with the trees,' Māret says sadly, even though the birch sucks up all the light and nutrients so the lichen the reindeer need cannot survive. 'I can always adapt to nature. Nature is always changing – we have always had to be prepared. I am hopeful of nature, but not humans.'

Māret's helpers spread reindeer skins on benches and set the tables. She goes back to slicing up meat as the first delegates begin to arrive in the hall dressed in their finely embroidered traditional felt jackets and reindeer-skin trousers and shoes. Hallgeir comes over to say hello and draws water from a large cooler into his own personal carved birch cup. I see that most delegates are carrying their own cups on a string attached to their belts.

'A big meeting! A very big meeting!' he says with a clandestine wink, as though he is a double agent, which in a sense he is.

Sámi representatives from all over the north of Norway have gathered to discuss the new reindeer law, proposed mining and wind farm developments in Finnmark and Tromsø and a climate change adaptation fund to help the Sámi transition to new livelihoods. But Māret sees the problem as much larger than Norway. 'Someone has to pay for this life, this lifestyle – and it seems it is the animals and our indigenous way of living. That is the cost.'

The Sámi identity, their oneness with nature, dies with the habitat that sustains it. They are feeling the brunt of climate change now, but in the long run it will be the people in hotter places or in coastal cities who will be in more serious trouble from floods and heatwaves. The Arctic is projected to come under increasing pressure from refugees as people flee crop failure and extreme temperatures further south.[4] The Sámi will probably be able to adapt where they are.

Māret is not smug, though; she is worried for everyone else. She wants people to see what is happening in Norway as a warning: 'You are not on the top. Nature is on top. And if you are working against the top, it will come and attack you. And that is what is happening.'

3. The Sleeping Bear

Dahurian larch, *Larix gmelinii*

Krasnoyarsk, Russia
56° 01' 00" N

Of all the forests in the world, the Russian taiga is the greatest. It covers over half of the land mass of Russia, stretching for more than three million square miles across two continents and ten time zones. It is a green carpet of trees atop permafrost and makes up well over half of the boreal's planetary engine, regulating patterns of wind, rainfall, climate and ocean circulation in the northern hemisphere. The boreal is largely made up of the Russian taiga, and the taiga is predominantly larch.

From the White Sea, underneath the chin of the horse's head of Fennoscandia where Norway meets Russia, all the way to the international dateline in the Bering Strait, the treeline and the northern reaches of the taiga are made up of endless stretches of one genus, the *Larix*. More than one third of the taiga (37 per cent) is larch, it is the keystone species. Like pine in Scotland, the larch has mastered its environment better than any other tree and therefore dominates the ecosystem, shaping the life cycles and evolutionary paths of other plants and animals. Its leaf litter is the basis of the soil, its cycles of seed production regulate the populations of birds and rodents, its demand for light sets the limits for what can grow in the understorey, and its resistance to fire as well as its appetite for colonising burned ground mean it has shaped the very architecture of the taiga forest in Siberia as we know it.

To consider the future of the boreal forest, you have to look at what is happening in the taiga. And the changes here hinge on how the larch is responding to the incredible warming taking place across Siberia. No one knows more about larch than the scientists of Russia's elite forest research institution, the V. N. Sukachev Institute of the Forest of the Siberian Division of the Russian Academy of Sciences, located in the city of Krasnoyarsk. The district of Krasnoyarsk is also the home of Ary Mas, the fabled forest at the top of the world, the most northerly trees on earth.

Krasnoyarsk is four time zones east of Moscow and slightly north of the point where the borders of Kazakhstan, Mongolia and Russia meet. It is home to one of the world's largest aluminium smelting facilities. Waking up in a backpacking hostel on my first morning in the city, weak February sunlight struggles to penetrate the haze hanging above the skyline. The room is sweltering from a huge radiator running the entire length of one wall. It is only minus ten, the receptionist told me when I checked in after my long flight from Moscow the night before. The heating was built for colder times. I open the two layers of window and immediately realise my mistake: icy air rushes in but it is not refreshing. It carries a nasty acrid tang.

The reason the coldest part of the year in Siberia, and the Arctic in general, is February rather than the middle of winter, when the days are shortest, is because the snow-covered ground continues to radiate heat out into the atmosphere, cooling the earth. Snow reflects short-wave radiation, insulating the ground beneath it from both the warming effect of the sun's rays and the cooling effect of the cold air, keeping the sub-nivean (under the snow) world a steady temperature. But while snow reflects short-wave radiation – the reds and yellows in the light spectrum – it absorbs long-wave radiation – the blues and greens – which at night it re-radiates back out into space. At night the coldest air is that closest to the surface of the snowpack. This is why if you sleep beneath a tree you'll be warmer, since the tree reflects the long-wave radiation back down to the surface instead of allowing it to escape upwards. This cooling effect of the snowpack creates an energy deficit, which in calm weather can result in a temperature inversion where the air at ground level is colder than the atmosphere above. In stable high-pressure situations for days or weeks on end, with the snow continuing to radiate heat away from the earth, temperature inversions of several hundred metres can develop, in which the normal process of hot air rising is arrested. In the clear air of the tundra, water vapour can become trapped and freeze as fog. In cities, air pollution that would normally rise instead becomes trapped lower down, and Krasnoyarsk, once declared by Anton Chekhov 'the best and most beautiful of all Siberian towns', is instead now famed for a hazy smog that cuts the lives of many of its residents short.[1]

Chekhov was mesmerised by the magnificence of the River Yenisei, and it is the river that defined the town then and does still. Siberia's third largest city was founded as part of Russia's expansion east in 1628 when Cossacks built a fort at the confluence of the Yenisei and Kacha Rivers against attacks by indigenous peoples. Just as Western European nations were sending out expeditions and convicts to new lands overseas in the seventeenth and eighteenth centuries, Russia was doing the same on land, exploiting the Siberian native peoples, according to Chekhov deploying 'the usual practice of getting them dependent on drink'. Serious colonisation of Siberia however had to wait for a road and the famous Krasnoyarsk bridge over the huge River Yenisei finished in 1741 that now adorns the Russian ten-rouble note. Development accelerated again after the discovery of gold and the construction of the Trans-Siberian Railway in 1895. Krasnoyarsk was where eight Decembrists were banished after the failure of their revolt against Tsar Nicholas I in 1825 and, continuing the tradition, constituted a hub of the Gulag work-camp system during the Stalinist era. There is still a penal colony to this day. During World War II numerous factories were relocated from the western Soviet Union to Krasnoyarsk, out of reach of invading German armies and close to the slave-labour workforce of the Gulag.

As the gateway to Siberia, under the Soviets Krasnoyarsk became a scientific and educational centre with institutes of metallurgy, aerospace, medicine, agriculture and technology. The largest and most obvious natural resource was the taiga, and so the first academic facility to be established was the Siberian Institute of the Forest, founded in 1930. After the war, this expanded to become the Siberian State Technological University, one of six universities in the city, and the Sukachev Institute, founded in Moscow in 1944 and relocated to Krasnoyarsk in 1959, assumed the mantle of Russia's premier forest research centre.

The institute is named for its founding director, Vladimir Sukachev, a larch specialist and a little-known pioneer of global ecology and environmentalism. Sukachev wrote a groundbreaking work of ecology, *Swamps: Their Formation, Development and Properties* (1926), that persuaded zealous Leninists bent on maximising agricultural production not to drain too many of them. 'Biogeocoenosis' was his broader definition of Karl Mobius's concept of an ecosystem encompassing the atmosphere as well as rocks, soil, plants and animals all engaged in a constantly developing interaction. He inspired a movement of student conservationists in the USSR a million strong in the 1950s, and he was the first to discover that larch will put out adventitious roots (from non-root tissue) in order to extract moisture from permafrost. This insight is perhaps his most important contribution to understanding the past evolution of the taiga and therefore its future. It was Sukachev who first understood that the larch's relationship with ice was the foundation of the Siberian landscape as we know it.

Each time an ice age brought a tide of glaciers down from the north, this obliterated the vegetation, forcing species into corners where the glaciers could not reach. Then, after each melt, the plants, trees, animals and latterly humans would emerge from their redoubts – refugia as scientists call them – and begin the process of colonisation again. Larch's strategic response to this long game of natural selection has been its readiness to hybridise, demonstrating extremely high ecological adaptability to diverse growing conditions from the Baltic to the Pacific and from the treeline in the far north all the way to the southern limit of the taiga at the 45th parallel in Mongolia.

The huge variety of adaptations in the size of cones, the number of seeds, the colour of needles and the amount of red and pink colour on the anthers of the trees' flowers provoked long-running squabbles and obsessive studies in Russian forestry for most of the twentieth century. Soviet researchers at the Sukachev Institute weighed seeds, counted needles, distilled the oils of multiple species and subspecies and tested propagation methods. One finding, that ground water no higher than 1.5 metres was a necessary prerequisite for successful seedling establishment, would be important for the future. One poor researcher, Abaimov, spent most of the 1980s measuring 30,000 larch cones in search of a metric for telling species apart only to discover that it was a useless exercise. The angle of the seed scales in each cone, as discovered by Sukachev in 1924, remains one of the only consistent distinctions between species, which are otherwise very hard to tell apart.

All larch have a noble air, their fine delicate needles, bright green in spring, lustrous orange in autumn, contrasting with a grey, somewhat scaly bark like that of the Scots pine. In their massed ranks they can blaze across a hillside. Larch is the only deciduous conifer. A corkscrew top allows every branch the maximum amount of light, each limb sweeping down in elegant curves that appear to drip with needles. In spring it puts out small purplish cones resembling sweets which slowly turn orange with the year.

There is general agreement now that the taiga is made up of four distinct bands of larch species (taxons), painted like vertical stripes across Siberia.[2] From the White Sea in the west to the Ural Mountains east of Moscow, the subspecies *Larix sukaczewii*, discovered and named by Sukachev, is dominant. From the Urals the versatile but permafrost-shy *Larix sibirica* (Siberian larch) picks up the baton until it meets *Larix gmelinii* (Dahurian larch) at the Yenisei, where the two species have hybridised to form a narrow strip eighty miles across of a hybrid called *Larix czekanowskii* in honour of the Polish botanist who discovered it. From there across the vast middle of Siberia known as the high north beginning at the Taimyr peninsula all the way to the Lena River, *L. gmelinii* is dominant, the undisputed king of the cold. A related subspecies, *Larix gmelinii var. japonica*, sometimes called *Larix cajanderi* (the Kuril larch), forms the eastern band of the forest from the Lena east to the Chukchi Sea and the Bering Strait, Kamchatka and the Sea of Sakhalin, the birthing ground of the great Pacific whales.

Sukachev hypothesised that the contact between these different species rubbing up against each other and hybridising was relatively recent, that they had all sprung from different refugia at the beginning of the Pleistocene. He showed that the dominance of Dahurian and Kuril larch went hand in hand with the permafrost. Indeed, as you go further and further into the interior of Siberia, these species form increasingly larger percentages of the forest, until in the far north pure stands of larch march across the landscape for thousands of kilometres, alone. Wherever there was ice, the Siberian larch and other subspecies gave way to the youngest, toughest, new kid on the block, the Dahurian. But now that the permafrost is melting, the balance of forces in Siberia is shifting. The destiny of the taiga and the permafrost have ever been linked, but until recently the foresters of the Sukachev Institute only looked up at the trees, not down, at the ground beneath their feet.

In the Soviet era it was common practice to cluster scientific institutions together in their own municipal area rather like a modern-day business park, either to foster innovative thinking by hothousing scientists or else to better control them. Krasnoyarsk's academic town, Akademgorodok, is about eight kilometres outside the city. A taxi, my hotel receptionist insists, is the only way; she's not willing to entrust me to the city's bus system.

Walking to the taxi stand, the town's industrial purpose is ever present, coating my nostrils in a black film. I cross wide parks planted with specimens of the boreal with paved pathways looping along a bluff above the River Yenisei. Even here, 3000 kilometres south of the Arctic coast where it joins the sea, the Yenisei is enormous, already nearly 2000 kilometres from its beginnings in Lake Baikal. The river is the heart of the city, mist swirling in figures of eight on the black water in its crusted icy frame. The river is liquid because of the hydroelectric plant thirty kilometres upstream, which prevents the water from freezing and provides most of the region's power. Far out on a frozen slab are hooded figures drilling holes in the ice and lowering lines in search of fish genetically adapted to wintering under ice.

On the other side of the park I walk along a grid of streets dotted with nineteenth-century wooden mansions with carved filigree facades. Another park has a children's playground sculpted from ice and a pole-mounted public address system pumping out muzak. The path is lined with birch, larch, pine and spruce beneath which are piles of shredded seed, birds ransacking the spruce cones for winter food. The central plaza containing the theatre, town hall and ballet is dominated at one end by an enormous intersection with eight lanes of traffic and a concrete system of pedestrian bridges where homeless people with no gloves hold out paper cups for change. At least it's only minus eighteen according to the digital display on the front of the theatre.

'Akademgorodok!' The taxi driver nods, and we lurch forward down a wide highway and out into the woods. Krasnoyarsk is in the middle of the taiga. Views of sharp-forested hills slicing away into the distance open out on all sides. The academic town is bordered on one side by a steep drop to the river over the high sandstone cliffs that give the city its name: *krasnyi* – 'red' and *yar* – 'cliff', and on the other by an impenetrable wall of forest. The air clears to reveal huge Scots pines glowing orange in the morning light while the silver stripes of the birch in between flicker like strobes. The larch begins further north, but before I visit the trees themselves I must talk to the scientists, to learn how to look at the landscape.

The changes in Siberia are complex. It is not a simple question of deforestation, as in Scotland, or afforestation, as in Norway, but a slow-motion transformation of the structure and composition of the forest. In some places the treeline is not moving at all, in others it is actually in retreat, while to the south the forest is burning up and not regrowing, and in the middle larch is giving way to other species less efficient at absorbing carbon and producing oxygen. And all the while the permafrost is melting, overshadowing everything else.

We pass through what was once a checkpoint into a grid of

large angular concrete buildings hunched in the snow with the feeling of an industrial zone. In the centre is a coal-fired power station puffing out yellow-brown smoke. Residential areas seem to be dotted about as an afterthought: apartment blocks with identical frozen children's playgrounds – iced swings hanging stiff in the windless morning air – mark every third or fourth street, but there are no people in sight.

The taxi drops me outside a nondescript concrete block similar to all the others, built in the 1950s.

'Sukachev?' I ask.

'Sukachev.' He nods and pockets the roll of roubles I offer without a backward glance.

On the steps in the bright sunshine, in a hat, gloves and light coat, Nadezhda Tchebakova is waiting for me. She has short grey hair and the quick but fast-fading smile of a humane scientist struggling to reconcile the knowledge she carries inside her with the madness of the world without; she laughs a lot but that doesn't mean she finds things funny. The generous research grants and high status of the Soviet era have passed, but the Russian culture of secrecy and suspicion is still alive and well. I am not allowed inside the institute without various letters and approvals that must be issued three months in advance so Nadezhda waves her hand dismissively and suggests we go for a walk instead.

Below, the black sweep of the river cuts a line between the cliffs covered in blinding snow. Above, a daytime half-moon lingers in the now visible cornflower-blue morning sky. On the far side of the valley smokestacks emit ribbons of black smoke, and red and black tower blocks throw shadows across the water. On this side, the gold dome of a brand-new Orthodox church gleams amid the trees, testament to the piety of an oligarch. Running down to the river from the institute is a forest inside a fence, the research arboretum, a collection of 400 species where apparently the oligarch would like to build dachas. Every boreal species and more is there, a whole crowd of glittering trees, in their heavy jackets of snow, needles sheathed in ice.

'I love it here,' says Nadezhda, taking in the view of the river, the rims of her spectacles catching the morning sun. 'Nature, animals, the forest, that's my thing.' But when she first arrived forty-seven years ago it was a very different story. She was depressed. She missed her family. Her parents had sacrificed a lot for her career only for it to take her away from them. They were farmers who had left the Volga during the hard Stalinist times of the 1930s and found work in an automobile plant nearer Moscow. Later, Nadezhda's mother was a cook in a kindergarten. They inspired her to study, and she ended up reading English at Moscow's best university. Her mother sent her entire salary to support Nadezhda, who failed her first year and made the fateful switch to study geography instead.

In the 1960s, ahead of the West, the Academy of Sciences in Moscow and Leningrad was the centre of cutting-edge research on what the soon to be world-famous climatologist Mikhail Budyko called 'the problem of climate modification by man'. In 1961 Budyko had presented his paper 'The Heat and Water Balance Theory of the Earth's Surface' to the third congress of the Geographical Society of the USSR, in which he stated that anthropogenic climate change was now inevitable and that human energy usage needed to be addressed. In 1962 he published the landmark article 'Climate Change and the Means of Its Transformation' in the USSR's *Bulletin of the Academy of Sciences*, explaining how the destruction of ice cover would cause 'a significant change in the regime of atmospheric circulation'. Nadezhda was newly arrived at the Academy of Sciences when Budyko published his 1969 article 'The Effect of Solar Radiation Variation on the Climate of the Earth', which explained how the polar sea ice/albedo feedback mechanism would drive climate change. She was beginning postgraduate studies when his seminal book *Climate and Life* came out. Once it was translated into English two years later in 1974, it defined the emerging field of climatology. By then Nadezhda was hooked.

When she arrived at the Sukachev Institute to continue her postgraduate work on climate impacts on boreal vegetation in the mountainous forests south of Krasnoyarsk, she was surprised to find that she was the only one doing climate modelling in a faculty full of foresters. Devoted to their summer ritual of field studies in the woods, they disdained models. Nearly half a century later, she is still the only one.

After Sukachev left, the institute was led by dendro-climatologists who spent much time looking back trying to reconstruct Siberia's past climate, but no one, it seemed, was interested in looking forward. International scientists became aware of Nadezhda's work, though, and in 1989 she was invited to Vienna to work at the International Institute for Applied Systems Analysis with top climate scientists from around the world. In Vienna Nadezhda worked on large models for global vegetation. She returned to Krasnoyarsk with a renewed sense of urgency and wonder at what was possible, and set about building future models for the forests of Siberia using the freshly minted scenarios from the Intergovernmental Panel on Climate Change. What she found was alarming. The northern treeline was projected to shift polewards slightly, but the real change was the taiga's southern limit. With increasing drought and more frequent fires, the steppe of Central Asia was projected to expand, consuming the burned-out taiga as it went, preventing it from regenerating: the greatest forest on earth would die from below.

'And that,' says Nadezhda, 'is what we are seeing now.'

In a warm cafe serving frothy hot chocolate and taiga tea – a mix of mint, willowherb and wild raspberries – I meet Elena Kukovskaya, Nadezhda's colleague and one of the foremost experts on wildfires in Siberia. While students tap away on laptops all around us, she explains that fire is a natural part of the forest cycle; indeed it is the reason the taiga has assumed its current form. Larch forests are relatively open with wide spaces between the trees and dense leaf litter which keeps the ground moist and prevents the build-up of the thick understorey that can fuel extreme fires. Larch with their deciduous needles and thick bark are fire resistant and long lived, provided the fires are not too severe. During our current era of the Holocene (the period since the last ice age), fire used to move through pure stands of larch forest with a light touch, scorching branches and trunks without killing the mature trees but consuming the leaf litter and top layer of soil to reveal the mineral soil below – the soil that larch seeds need to germinate. The forests of North America meanwhile are generally much younger, less than 200 years old, as fires will destroy whole stands of spruce, aspen and pine, meaning cohorts of seedlings establish at the same time, resulting in what foresters call even-aged stands. But fire dynamics are changing. In recent fires that Elena has studied, trees are not coming back at all.

In the high taiga at the treeline, in Ary Mas for example in the north of the district, fire was always a rare visitor, usually the result of lightning. There the interval between fires is long, up to 300 years, but frequency and intensity increase as you go south. In the lower latitudes the fire interval was previously 5–30 years depending on rainfall. Now in some places fires are an annual event. As temperatures have increased and the soil has grown drier, fires are hotter, longer and more frequent, consuming more of the soil and making it harder for larch to re-establish afterwards.

'People have tried planting, but then they get burned the next year too,' she says.

Repeated burning makes terrain almost impossible for trees to root in. Instead plants that like very marginal soils have taken their place. This means shrubs like willow crowding out the larch and building up dense clusters of wood that in turn burn hotter next time. Over time, the burn cycles open the door to grasses from the steppe, which prevent trees from germinating and smother everything else.

Further north, hotter fires and drier summers mean that in the middle taiga, north of Krasnoyarsk, Scots pine is replacing larch on burn sites. This has major implications for the structure of the forest and for how other scientists quantify and model its 'ecosystem services'. Since different species have different plumbing, their contribution to the atmosphere and climate is varied and unique. Of the carbon dioxide that the taiga sequesters, larch is responsible for soaking up 55 per cent even though it only accounts for just under 40 per cent of the trees. This single species is the greatest arboreal source of oxygen on earth. Because larch is deciduous, it transpires a lot more water than evergreens, taking in 20 per cent more carbon dioxide than pine, and the soils beneath larch, covered as they are with half-decomposed larch needles, emit a quarter less carbon dioxide than those under pine. Moreover, a warming forest is less efficient at cycling and sequestering carbon as trees lack water to photosynthesise, stop growing or lose their leaves earlier.[3] And yet so many global models rely on fixed assumptions about the amount of carbon forests can store and ignore how much they are changing. One study suggests by 2040 global forests will absorb half the amount of carbon dioxide they do at present, if current warming trends persist.[4]

An even bigger worry, Elena says, is whether there will be any forest left at all. The sheer volume that is burning is what bothers her now. The forest is drier. Unlicensed logging to fuel China's construction boom means there is much flammable waste. In addition, a warmer, more humid atmosphere means double the amount of lightning and double the number of ignitions. Elena's interest is no longer purely academic. She has just given birth to her second child. In 2019 black smoke covered the town of Krasnoyarsk for weeks, and people couldn't breathe.

'Every year, every fire season now, we wait, we brace ourselves.'

Over fifteen million hectares burned in 2019, an area larger than Austria. Prior to 2018 the average volume of wildfire emissions was 2 megatonnes of carbon dioxide per year. In 2019 it was 5 megatonnes. Later in 2020, after I visited her, fires in Siberia would break records, emitting 16 megatonnes in the month of June alone. Forest fires on this scale had not been predicted until 2060.

When I join her for lunch at the House of Scientists, Nadezhda smiles as I recount Elena's observations. She predicted these changes with her models a decade ago. She gives me a map. It is a printout from her latest published research in the US journal *Environmental Research Letters*.[5] What she is interested in now is what the ecosystem changes mean for humans. The article, later featured in the *New York Times*, has a technical-sounding name ('Assessing landscape potential for human sustainability and "attractiveness" across Asian Russia in a warmer 21st century') but dramatic implications.[6]

Nadezhda's previous work showed that even under moderate warming scenarios Siberia would continue to experience much higher rates of regional heating and that the forest will forge north if it doesn't burn first. But, given that it doesn't have much land to colonise to the north, this ultimately means that 50 per cent or more of the forest will be converted to steppe by century's end.

In the 1930s the USSR developed a system for classifying human living conditions based on climate severity and comfort. The system was used to compensate for harsh living conditions with salary bonuses and was the basis for successive bureaucratic efforts to encourage migration to Russia's east, subsidised by the state. There are seven categories, and the two best zones, 'favourable' and 'most favourable', are currently not present in Siberia, but Nadezhda's model showed that this is soon to change.

The lines on the map indicate that at the higher end of predictions – up to nine degrees Celsius of warming for north-eastern Siberia – the southern boundary of the forest will shift north by 1000 kilometres, allowing up to 85 per cent of Siberia, three million square kilometres, to become suitable for agriculture by the end of the century. Changes are already afoot: Russian agribusiness corporation RusAgro is expanding its extensive wheat farms near Vladivostok and immigrants from China are being encouraged to farm tracts across the Amur River. Meanwhile, North American yields in 2019 and 2020 showed their first signs of faltering. [7]

Under Nadezhda's latest model up to half of Siberia will become 'favourable' or above for human settlement, but Nadezhda and her co-authors stop short of the inevitable political conclusion. For most of human history humans have inhabited a very narrow band of land within a particular temperature range across the planet. But some parts of that band are already hitting the upper limits of that range, and by 2070 more than three billion people will be living outside it – most of them in South East Asia, only a border or two away from Russia's vast, newly fertile, east.[8] Within decades, parts of the Middle East and South East Asia will become too hot for humans to safely work outside or to sleep without air conditioning. Siberia is the obvious refuge for the large populations further south already under pressure from heat stress, flooding, drought and famine: the climate-stressed areas of eastern China, Bangladesh, Pakistan, Nepal, Uzbekistan, Kazakhstan and the Middle East.

The niche for trees and for human habitation is very similar. The same dynamic of expanding desert and steppe at the expense of forest is already under way in South and Central America and across the Sahel in Africa. If one were to model human migration in the same way as for trees, without regard for political borders, you would expect a massive shift north.

This link between the niches for human habitation and trees is something I wanted to know more about. Human society has become decoupled from its environment only recently – in the latter half of the twentieth century – with supply chains to remote and inhospitable places made possible by fossil fuels. But before that, apart from the Inuit with their source of energy from the blubber of whales and seals, humans have never managed for long without a source of wood. In defining ecosystems and habitats, the treeline has shaped the possibilities for human existence and, by extension, set the terms for human culture. Our place has always been at the edge of the forest, with a relationship to it.

One of the reasons I am so keen to visit the larch treeline at Ary Mas in the northern part of the district is to see this relationship in action among the Nganasan, a unique indigenous group who have lived above the Siberian treeline for thousands of years, overwintering in the world's most northerly forest, while spending the brief summer months hunting reindeer on the tundra closer to the North Pole. I ask Nadezdha if there are anthropologists among her colleagues at the institute, but the answer is no, they only study trees. She introduces me to one of them, a larch specialist called Aleksandr Bondarev who has been visiting Ary Mas for decades, and we arrange to speak on the phone.

For the most part, Nadezhda barely interacts with her colleagues.

Like on all campuses, most researchers are deeply engaged in their own work and oblivious of others down the hall. Nadezhda works day and night in her lab and again when she goes home to the apartment on the campus where she has lived for most of her life, alone except for her many cats. There is an urgency to her mission, perhaps partly because the rest of society is so complacent.

'There are no young students coming up in modelling of vegetative change. They are not interested. Maybe I am a bad teacher. I am impatient,' she says matter of factly. 'This research will stop with me.'

Modelling must be lonely work, peering into the crystal ball of sophisticated climate models for a glimpse of the apocalypse. And the emotional toll must be significant, although Nadezdha, like a good stoical Russian or a properly disinterested scientist, does not allow emotion in, even though her name means 'hope'.

'I do not have feelings about the future. I can't change things. I am just warning people. It is not my responsibility to solve,' she says without smiling.

When we discuss other researchers whose work might be relevant to my project – to the future – it seems apposite that the person she suggests to invite to lunch, a distant acquaintance of fifty years, is also, in his own way, asking the same question: where will all the people go?

Professor Dr Alexander Tikhomirov is a quiet, careful man. He joins us in the staff restaurant at the social club, where he eats slowly, dabbing the borscht from his clean-shaven chin. He has pure white hair, bushy eyebrows and small grey eyes that glimmer with wry humour. Dressed in a sweater, shirt and slacks, he keeps his black military boots firmly planted on the floor as we talk.

Like Nadezhda, Alex started his career in Krasnoyarsk about fifty years ago. He began in the Institute of the Forest researching how trees heal themselves after injury by insects, wind or over-browsing from animals; he was interested in how cells restructured. Then he switched to cultivated plants and was recruited into the Institute of Biophysics to work on a top-secret project called

Bios-3 in the Department of Life Support Systems, which he now heads within the Siberian Aerospace University. Bios-3 was concerned with human survival in space for extended periods – either in space stations or for a projected mission to Mars. It is what is called a closed ecosystem experiment. A bit like playing with climate change in miniature.

In 1972–3 three cosmonauts spent 180 days in a sealed chamber in Bios-3. They ate wheat and vegetables from an algal cultivator with xenon lamps approximating sunlight, and each relied on 85 square metres of exposed chlorella algae to produce their oxygen. The entire facility was 315 cubic metres and incorporated three cabins, a galley, toilet and control room. Human waste was dried and stored, but meat and water were imported. The Bios experiment achieved about 85 per cent efficiency. A later similar experiment in Arizona, in the United States, conducted by a private company called Biospace Ventures, claimed to have achieved 100 per cent efficiency with no extra water and oxygen required. The Arizona experiment collapsed however, when, following a dispute, a manager called Steve Bannon (later chief of staff to President Donald Trump) broke into the sealed chamber, endangering the astronauts inside. Much of the research data was lost. The facility was taken over by Columbia University and then the University of Arizona. It is still used for research on habitat modelling. Among other projects, a half-acre of rainforest is growing in a glass pyramid at extremely high temperatures, suggesting that, with enough water, some forests might be able to survive a hot-house earth if the carbon dioxide fraction remains constant, a big if.[9]

A different joint experiment between Bios-3, the Arizona facility and the European Space Agency is contributing vital information on what happens to plants and humans in a carbon dioxide-rich environment. This research has shown how oceans and coral reefs will be devastated by acidification and, importantly, that there is a carbon dioxide saturation point for some ecosystems. Plants, it seems, cannot cope with ever increasing levels of carbon dioxide. Photosynthesis depends on a balance to work; if plants don't get enough water and light to make use of the carbon dioxide available it can overwhelm them. Some species are better adapted than others of course; the prehistoric ferns and ginkgos that remember the Carboniferous period may find distant clues in their genes that will help. As for humans, Alex's team found that an atmosphere of more than 1 per cent carbon dioxide (10,000 parts per million) was obviously deleterious to human function – the cosmonauts became confused and uncoordinated.

Alex will not be putting any more humans into a controlled environment until they have fixed various problems with Bios-3. For one, maintaining the atmospheric balance without storing or expelling human waste is very hard. Methane, ammonia and other products of decomposition are very bad for plants at high concentrations – they make them age faster.

'And humans too!' says Alex with his signature dry chuckle.

Methane is his obsession at present. His team is using the closed chambers to do artificial warming.

'We want to know the critical temperature when methane accelerates global warming. That is a key question!'

He gets agitated when talking about the details of the experiment: how his colleagues are transporting and freezing blocks of tundra then gradually cooking them in the herbarium and capturing the gas, but when I ask him what it means he is quiet.

'I have grandchildren,' he says, suddenly sombre. 'I want the best for them to be happy, but . . .' He looks up at the yellow ceiling then out to the frosted campus sparkling in the early afternoon light. 'I see three phases. One, maybe warming will bring benefits. That phase is gone. Two, flora and fauna will shift. That is where we are now. Three, humans will have to adapt to new conditions – soil, agriculture, fight for land, water, resources. And if you consider nuclear weapons then . . . the picture is frightening.'

Outside the window a gang of waxwings is loudly plundering a rowan tree.

'It is an urgent priority to try and reach Mars. If we had a global effort, we could get there in twenty years. Install a station. It depends how it will go on Earth.'

*

After lunch Nadezhda walks me to the bus stop. Our lunchtime conversation seems to have unlocked a franker self. As if, emboldened by Alex, she can now speak her mind.

'It is my belief that a chain reaction has already started. The permafrost is already melting. It is hard to see how it can stop.' She pulls her thin coat about her. She didn't replace her old thick winter coat; winters are not cold enough to merit it any more and Nadezhda is a frugal person. She wishes me luck in Ary Mas and recommends I search for information about the Nganasan in Krasnoyarsk Regional Museum and also that I go to the ballet in the city, 'just because'.

At the bus stop she is reluctant to wait too long. 'I've got a lot of work to do, I'm sure you understand.'

Out of the windows of the bus on the ride back to the town centre the low afternoon light flickering through the ghostly frosted birch forest assumes the black and white stripes of a Siberian tiger before the billboards and ruined concrete of the outskirts of the city take over. At the terminus I follow Nadezdha's advice and seek out the Krasnoyarsk Regional Museum. It turns out to be the most appropriate place to search for information on the Nganasan for several reasons. The Nganasan language and culture is almost extinct, so what remains is, literally, here in the museum. The building also faces the famous bridge that opened up Siberia to the Cossack invaders, the very cause of the indigenous people's demise.

The museum is warm and extravagantly staffed, and I am pleased to be inside as darkness creeps over the frozen, polluted city. The elaborate concrete Egyptian facade that frames the huge hardwood doors of the entrance has little relationship to the indigenous treasures inside collected from Siberia's endangered original inhabitants. In a curious hierarchy, the large wooden boat that occupies the entire ground floor opposite the entrance is a *koch*, a vessel with a reinforced hull built by the tsarist colonists for exploring the ice-bound Arctic coast. The communist era is represented by an exhibition on the galleried first floor, with images of smiling collective workers and a history of space exploration, complete with a satellite hanging from the ceiling in homage to Krasnoy-arsk's heritage as a launch site for rockets. Prehistory and the indigenous cultures are confined to the windowless basement.

In the dimly lit room a series of pen and ink drawings explains that the oldest evidence of human settlement in Siberia dates to 70,000 years ago. The last ice age, what in Siberia was called the Sartansk glaciation, covered much of the terrain for 20,000 years between 50 and 20,000 years ago. Since then the retreating ice has revealed new lands into which mosses, lichens, grasses, shrubs and trees have swiftly moved. Animals followed the grazing and then humans followed the animals. Most of the Siberian peoples share a heritage with the pre-Samoyed Neolithic peoples known as the Yukagir, who emerged from their glacial refuge in the mountains of north-east China in pursuit of the ever-widening ranges of the migrating reindeer. They spread all across the continent of Asia, from the Pacific to the Ural Mountains, very likely crossing the land bridge of Beringia into Alaska. In one direction, linguistic analysis has found links between Yukagir and the Finno-Ugraic languages of the Baltic states and Hungary. And in the other, 'birch' is one of thirty-six words shared between the endangered Ket language of Siberia and the Athabaskan Na-Dene group of northern Canada.

The indigenous artefacts in the museum are remarkably similar across the groups: cylindrical nomad tents, reindeer skins sewn into clothes decorated with beads, metal ornaments and dye and hunting gear, canoes and snowshoes made of larch and birch. The Selkup, Ket, Evenks, Nenets, Enets, Dolgans, Khakas and Nganasan occupied a huge range of territory and environments but they shared a very similar animist belief system. 'Saitan', a Dolgan word, meant a spirit that inhabited an object. Spirits were a central organising principle for the people of the taiga. The authority responsible for mediating the relationship between the spirit world and the 'upper' world of the humans in every culture was the shaman. The taiga peoples believed that trees were antennae, crucial to communication between the upper world and the lower.

In the locked glass cases are the mournful confiscated sacred objects of the shamans: round skin drums decorated with images of reindeer, birds, humans, the sun, moon and stars. The drum was a boat to be sailed in, a reindeer to ride in, a vehicle for the shaman to journey far, to see what was happening elsewhere and report back. His utensils would usually be made of larch – the cleanest tree, free from spirits. The drum symbolised the circular nature of the universe. Along with unique songs and poems, a strange music inspired by the sounds of nature, of birds, wind, water and trees, with strict rules of metre and rhyme, the drums were used to summon spirits to assist in the trials of the upper world.

A Korean law professor is the only other visitor in the basement. He explains that the roots of Korean law come from animism and Buddhism, both religions of the forest. Korea too is a boreal nation, he says.

The shamans of Siberia were systematically persecuted by the secular Soviet state, murdered, imprisoned and, in one story, thrown out of helicopters and told to fly. The belief systems that remain today are for the most part scraps of stories preserved by a few, soon to pass out of the limit of human recall. When they crossed the River Ob and then the Yenisei into the heart of Siberia, the exploring Russians disrupted a rich web of cultures of the forest that is in precipitous decline. Many groups have been assimilated, others have almost died out. There are almost no Enets people left. The age of extinction is not just a matter of animals and plants, but ethnicities, languages and cultures too – many have already gone.

The Nganasan avoided Soviet control for the longest. They retreated to their frozen redoubt in the forest of Ary Mas on the Taimyr peninsula, where they worshipped the tree lord who lived in the larch. The fiercest of the wild tribes, with the most powerful shamans, until the 1930s they were actually unknown to the Soviet authorities. The government called them Samoyeds. The Nganasan refer to themselves as *nanuo nanasa* – 'real man', a name that suggests an acknowledgement of difference.

Apart from occasional attempts to elicit fur tribute from the 'Pyasina Samoyeds' and a handful of brief descriptions in nineteenth-century works, the Nganasan do not appear in the written record of Russia or the USSR until 1936, when A. A. Popov, a young ethnologist, was commissioned by the Institute of Ethnography of the USSR Academy of Sciences to study them. Popov spent two years living with the nomads of Taimyr, the most northerly indigenous peoples of the world, travelling more than 6000 miles over the tundra to the north of Ary Mas up to the 75th parallel, learning their language and customs, taking 800 photographs and collecting 500 objects.

The people Popov encountered were practising an entirely self-sufficient nomadic life hunting reindeer far above the Arctic Circle just as their ancestors had for thousands of years. He met women who rolled reindeer sinews against their cheek to form thread, sewed parkas and trousers from reindeer hide and made the soles of boots from the skin of reindeer foreheads. They used the animals' thick winter fur for winter clothes and their thin skin for summer garments. And he met tireless, strong men who hunted with bows and arrows made from the stripped roots of larch, wrapped and strengthened with birch bark and waterproofed with fish glue.

He hunted with them for two seasons, on the road for several days with only tea, a kettle, skins and a sled, witnessing mass reindeer hunts, collective slaughters of hundreds of animals at the same place each year on the reindeer's migration routes. He describes hunters staking out lanes miles long with white ptarmigan wings attached to the top of poles to corral the deer into a funnel, either a narrow gorge or a cliff, leading into a lake where hunters in canoes would stab the animals with short spears in the rear so as not to spoil the hides. They would tie several carcasses together and drag them to the shore. Then 'these people who do not know fatigue' would dress the kill, render the fat, viscera and skins before repairing their hunting gear and sleeping for two or three hours and then repeating the process. Mass slaughter, Popov said, was much easier than hunting single reindeer on foot when one had to carry the dead animal back alone.

The fat and meat would be dried and smoked, the brains and

other viscera eaten raw. Before winter, people would eat until they were nauseous to stock up for the long cold, at least 263 days below freezing. By spring they were often starving until the migrating birds returned. Then the Nganasan used decoys, nets and bows and arrows to kill ducks and geese, up to a thousand in one go, the skins, feathers and meat all put to use immediately without regard for sleep and the rendered fat stored in reindeer stomachs. Larch firewood was preserved for smoking and curing. Food was eaten raw if possible, and in their fur-lined tents they stopped lighting fires from March onwards to conserve fuel, huddling together with their dogs to keep warm.

But Popov's visit was the last glimpse of this ancient way of life. In 1938 he wrote proudly of the impact his fieldwork had had:

> Before the great October socialist revolution the Nganasan were one of the most neglected small peoples of the Siberian north and were doomed to extinction. Lost on the vast tundra, they were isolated from the outside world. Elements of civilisation practically did not penetrate into their way of life. [Now] . . . Nganasan children study in schools which have opened in distant settlements. The first railroad has appeared on the tundra, and barges and steamboats have appeared on the rivers. Agriculture is developing, and vegetables now grow above the Arctic Circle.[10]

Popov did his job too well. It was not 'neglect' that 'doomed' them to extinction, but the state's attention. Colonisation, whether socialist or capitalist, is so effective, so brutal at alienating people from nature that in one lifetime a culture that evolved in relation to the treeline over millennia has evaporated almost without trace.

According to the 2012 census there were fewer than 500 Nganasan left. But many of them no longer speak their language or practise their culture and all have been removed from their ancestral ranges to live in towns far away. There are only a few extended families remaining along the River Khatanga on the Taimyr peninsula, over one hundred kilometres south of Ary Mas, way below their historical territory. For a hefty fee, a bespoke travel company has arranged to supply me with a specialised Arctic truck and a

translator so I can attempt to interview them and visit the cryolithic larch forest that they called home for over 8000 years. And
so the following morning I have a 4 a.m. appointment at the airport with my contracted translator, Dmitry, for the next stage of
my journey to the far north.

Still, I cannot ignore Nadezhda's urging to see the Krasnoyarsk
ballet. The museum is just across the main plaza from the State
Theatre. High on the north bank of the river, this looks out across
the string of red lights hanging from the bridge spilling their reflec-
tions over the black water. With smiles and sign language through
a thick window, a woman in a fox-fur hat and glasses manages to
communicate that there are tickets available. I take my place
among the city's furred and booted middle classes, on pale blue
velour seats in the domed modernist concrete theatre for a full
orchestral version of *Swan Lake* in which both the white swan and
the black swan are danced by men. I am no judge of ballet, but by
the end I am just as moved by the rest of the audience to leap to my
feet clapping madly not so much in appreciation of the perform-
ance, though it is a rare delight, but in wonder at the achievement
of sustaining such culture in such an unforgiving place.

Uchukhtai – the Dreaming Lake – Russia
73° 08' 81" N

Siberia is so vast, it is the same distance from Moscow to
Krasnoyarsk – 3000 kilometres and five hours of flying across four
time zones – as it is from the south of Krasnoyarsk district on the
border with Mongolia to the Arctic port town of Khatanga in the
north. Dmitry – to me now Dima – and I fly for what seems an
eternity across the lightless expanse of northern central Siberia.
All the front seats have been stuffed with cargo, and everyone
sleeps until, in the final moments of the flight, the taiga emerges
below: swirling patterns of a frozen river system doubling back on

itself, tiny trees dotting the riverbank. On the runway we squint in
the weak light of a hazy mid-morning dawn at an aviation grave-
yard of prop planes, jets and an Mi-26 helicopter encased in snow
and ice.

Beneath a larch tree blasted with ice we meet our fixer Alexei,
his grinning face framed by a fox-fur hat with ear flaps, astride a
snowmobile towing a steel cage for our luggage. The cold is, literally, blinding. At minus forty-four the tiniest breath of wind makes
your eyes water; tears freeze on the skin, and if you blink too long
your eyelids stick together. The icy air abrades your throat like
sandpaper. The cold penetrates your clothes like needles, and you
cannot be without gloves for more than sixty seconds before the
skin starts smarting as if on fire. It is normal to wear two hats, two
coats, two pairs of trousers and socks and double gloves, either
made of fur or stuffed with down.

At speed, through viscous, watery eyes, I glimpse Khatanga, a
medium-sized town with large square buildings. There are rows of
pale green prefab blocks and one red-brick apartment building all
connected by enormous pipes lagged in shiny aluminium foil tra-
versed by bridges and walkways: the municipal hot-water system.
The skyline is marked on one side by derricks hanging idle over the
river port and on the other by a power plant topped by two thin
chimneys with flashing red lights like lit cigarettes streaming grey
smoke into a yellow nicotine sky. Everything is covered in a fine
grey snow, like ash. But we do not linger.

Alexei delivers us straight into the hands of our drivers Kolya
and Kolya and their pride and joy, a giant white truck called a
Trekol with enormous tyres the same height as me. The engine is
running; a huge steel drum full of diesel sits on a trailer behind,
and a large spherical spare tyre is being strapped to the roof. One
Kolya has sandy hair and the other black. Both smoke cigarettes
clamped into wry smiles. Blond Kolya motions with a cigarette in
his oily ungloved hand for us to go inside the garage.

The door is a battered steel sheet blasted white, suggesting a
relentless polar winter of endless dark and storms. Three large fish,
frozen solid, each about as long as my arm, have been casually

tossed by the door, resting amid machinery and a pair of frozen reindeer hearts sitting on a pad of snow. An iron stove glows at one end, but its influence is limited. Oily benches, tarps, tyres and tools are piled all around in the gloom; there is one grimy bulb and no windows.

Kolya and Kolya come in with much bluster, shaking of hands and laughing about things I don't understand. Dima's translation is on a need-to-know basis. Alexei and Dima have a long conversation and then Alexei shakes my hand and says goodbye. Soon we are clambering into the back of the Trekol with Kolya and Kolya in the two upholstered front seats, and Russian pop music in the speakers. The spanner gear stick is pushed into first and the enormous truck gives a jolt and inches forward.

Trekols were designed specifically for Siberia with monstrous inflatable wheels to cope with the swamps of summer and the snow and ice of winter. They are very highly geared and creep tank-like over the landscape, rarely exceeding thirty kilometres an hour. We pass an abandoned geologists' camp on the outskirts of town and a coal mine where, with a bump, the Trekol leaves the road, whining, crunching and lurching over the frozen surface of the Khatanga River, travelling east towards the Laptev Sea, part of the Arctic Ocean.

Our first stop, Dima explains, is Novoribyne, along the Khatanga, one of the two most northerly settlements in Russia, where an extended family of Nganasan live. After that we will head north to the forest of Ary Mas and then visit the Dolgan nomads, who still practise reindeer herding above the treeline and have their own relationship to the forest.

'How far is Novoribyne?' I ask.

'A hundred and sixty kilometres.'

For the next eight hours Dima and I face each other on parallel bench seats in the back, our bags and a jerrycan of vodka stashed between us, clinging on for dear life, tossed about as if aboard a ship in a squall.

I try and focus on the pitching and rolling view, but the sun, which has only just returned after the dark polar winter, sets behind us as soon as we leave the town in a single yellow line across the horizon, just before 3 p.m. In the twilight I make out through the frosted windscreen a wide expanse of flat white river with low hills on either side. We are tracking the treeline. There is a fringe of larch trees running along the southern shore but none to the north. On that side of the river is the Taimyr peninsula, a bulb of land 1000 kilometres long, the northern tip of Siberia dividing the Kara Sea from the Laptev, almost exactly halfway between Norway and Alaska. Thousands of kilometres due south is Lake Baikal, and further south still on the same longitude are the cities of Ulan Bator, Hong Kong and Jakarta. Taimyr is as close as you can get to the North Pole on any continental land mass; only the islands of the Arctic Ocean are nearer. Were it not for the strange patch of trees at Ary Mas – the name means 'tree island' in the Dolgan language – this southern shore of the Khatanga would be the most northerly treeline in the world.

The forest, the tundra, the river, the sky are all just bands of different shades of grey in the gathering night. At some point blond Kolya switches on the roving spotlight on a steel pole that pokes out of the roof and I see the road ahead: the river seems to have frozen in the middle of a storm, great folds of ice and snow resembling waves. The Trekol, whining incessantly, rising and falling and bumping over them, ensures we do not sleep.

Around 11 p.m. lights appear out of the murk as if the river has unfrozen and we are viewing a port seen from the water. The relief is similar. The Trekol leaves the river and climbs a steep snowy bank cut with the tracks of other vehicles. Lights on poles reveal a collection of buildings with pitched roofs groaning under metres of snow and coated in ice. When the very air outside is dangerous houses are not just homes but refuges, sanctuaries, space stations, essential for survival. They lie deep under the snow but prone with the promise of warmth. It seems incredible, but Novoribyne is a village of several hundred people, with a church, a shop and a school with over one hundred children.

At the junction where the rise meets a row of several houses in an approximation of a street, the Trekol stalls. Kolya and Kolya

erupt in a flurry of what I imagine are swear words and leap out of
the truck. They only have a short window in which to restart the
engine before the fuel lines freeze. If that happens, they must light
a bonfire under the engine – a common but risky Siberian tactic.
Dima and I put on our snowsuits, gloves and hats and climb down
onto the street. Black Kolya points wildly at the nearest house,
shouting instructions to Dima that are lost on the wind. We walk
beside what seem to be garages buried in the snow with wires
strung between them coated in ice towards a flight of steps chopped
out of the drifts leading to a wooden house and a doorway in a
pool of yellow light flickering in the blizzard.

A thin covered porch, the walls furred with ice, holds a huge
cache of frozen reindeer carcasses. We knock on the door and are
soon in the kitchen of a genial man called Konstantin wearing a
white singlet and joggers. He darts around the house carrying children, cigarettes and a huge frozen fish which he places on the
vinyl-covered table and begins to slice into long icy strips – *kyspyt*,
a local delicacy. We eat it with tea, mustard, salt and a small dish
of chilli powder. Delicious!

An hour later the Kolyas return, triumphant. The Trekol is outside, purring obediently. Anna, Konstantin's wife, is at the stove
now, in jeans and floral slippers, talking non-stop to the visitors
while frying fish in a pan that she manoeuvres on the electric hob
with pliers. Before long, another meal appears: fried Arctic char with
bread and more tea. It is after midnight and still no news on where
we are spending the night. Dima and I have been awake since
4 a.m. Konstantin's little boy comes and goes with no sign of going
to bed.

After the meal, all the smokers including Anna and Konstantin
sit on the floor by the ceramic stove in the main sitting room. Two
salmon are thawing on top in a dish. Half an hour passes during
which they look at their phones, the GPS and talk. Eventually
Dima explains: the Nganasan family have a funeral, a family matter. It has been decided we will proceed to the next town, Syndassko,
where the Dolgans are, and visit the Nganasan and the forest on
our return leg to Khatanga.

'Tomorrow?' I ask, weak with fatigue.

'No. Now,' says Dima. I don't understand the Siberian way of doing things, clearly.

'Syndassko!' says blond Kolya, standing up, stretching and indicating the roof with his eyes.

'How far is Syndassko?'

'One hundred forty.' My heart sinks. The Kolyas, meanwhile, seem to be relishing the hardship, as if some elemental truth about the world is being demonstrated for the benefit of the pampered Westerner. Day and night mean nothing this far north.

Konstantin and Anna tell Kolya and Kolya about the perils of the sea ice. Keep close to the shore; don't go too far out. There are folds and ridges that are too high to drive over; you might get stuck. The sea froze badly last winter, plus there are cracks where upwellings of briny slush have caused folds and ridges to slide over each other. The cracks are a new phenomenon.

'OK!' shouts blond Kolya, and the Trekol lurches forward into the dark, back down the slope onto the frozen river once more.

Some time in the night, after several hours of whining forward into a blizzard, the Trekol stops. We get out to pee. The moon is lost in the murk, the shoreline is nowhere to be seen, there is only fog, the snow a few metres in front of the truck. At Syndassko the Khatanga joins the Laptev Sea in a huge bay, the Gulf of Khatanga, fifty kilometres wide. We are somewhere out there. We could be in space.

Where it appears beneath the shifting snow, the sea is like black glass, with an opaque and cracked glaze sometimes crushed into folds resembling tectonic plates. The Kolyas spend a long time peering at the GPS. I am worried. The two-ton truck is idling on perhaps a metre or two of frozen sea, miles from the shore. A Kolya jumps up and down laughing as if testing the ice. Until recently it was taken for granted that you could always drive on the sea for half the year. But during an expedition to the North Pole this winter scientists took to wearing lifejackets after several of them went through the brittle ice. Even the most solid certainties are suddenly liquid.

'Are we lost?' I ask.

'We are at the North Pole!' jokes Dima. Indeed, it is not far away. Three days driving even at this speed and we'd be there – if the ice held.

The sky is lightening when we finally pull into the port village of Syndassko. In daylight, the scale of the Gulf of Khatanga becomes apparent. Along the horizon, white hills are just visible – the edge of Taimyr on the far shore of the vast, frozen sea. All along the southern bank of the estuary are heaps of logs several storeys high. There are whole trees piled up as if a giant bulldozer had cleared a forest and dumped it here. They are all larch from the forest far upriver, brought down by the flood each summer when the ice goes out – historically around 20 July but increasingly earlier than that. We left the forest behind during the night at Novoribyne, now a hundred miles south, but the forest far inland continues to influence human geography many miles downstream: the driftwood caught in the sweep of the bay where the river makes its turn towards the sea is the reason the town of Syndassko is here. It is one of very few settlements above the treeline that is not a military or mining station.

Behind a shingle berm, a collection of houses half buried in snow and blasted with ice emerges from the night. We drive down a kind of high street that feels like it belongs in a Wild West town, a wide avenue with wooden frame houses on either side, wires heavy with snow connecting them. But it is hard to distinguish buildings, cars and oil tanks from each other; they are all just humps in the snow. We stop outside one of the houses and are soon seated around a tiny breakfast table with a man called Sergei. He throws a block of ice into a kettle, slices strips off a frozen fish and lights cigarette after cigarette as he pumps the Kolyas for information from Khatanga. We have come in from the cold, and must be fed. He is giving us what we need and seeking news from Khatanga in return.

Sergei has a ready chuckle, a round shining face and a handsome belly that he rubs as he talks. Everyone is a bit the worse for wear today, he explains. The helicopter comes on Thursdays, bringing supplies, among them vodka. The going rate for a litre is twenty-five kilos of frozen fish, and there are plenty of fish. Sergei wants to know about the cracks in the sea ice, did we see any? This is a new peril. The weather today is so warm! It has risen to minus twenty-seven overnight. The men all nod into their tea. Minus thirty was unheard of in February until recently.

It is still cold enough though to make going to the toilet outside a hasty affair. In the vestibule of Sergei's house is a lean-to full of large blocks of ice half a metre square, and on the other side a box of what look like huge slabs of coal. None of the homes this far north have running water, as the pipes would burst. Each house is ringed instead with a scattered pattern of yellow snow and brown human faeces. There is no attempt to confine it to one place. It will stay there, frozen, until summer.

Hearing that I am interested in trees, when I come back Sergei proudly shows me the 'fossilised trees' in his fuel basket, the remains of a forest, like Ary Mas, that was once much further north. The young black coal comes out of a cleft in the mountain about five kilometres south of town. Syndassko is the most northerly settlement in Russia. It was originally the site of a summer fishing camp of the Dolgan nomads; in winter they would migrate south, back to the treeline at the Papigai River. Then until the 1950s it was a trading outpost where nomads brought fur and reindeer to trade with the 'Russians'. There was one shop serviced by boat and no permanent houses until Syndassko joined the Gulag 'archipelago' after World War II. It was the prisoners who discovered the coal and showed the Dolgan people what it could do, opening up the possibility of year-round occupation of the place. Fossil fuels – prehistoric trees – were still the primary determinant of where humans could live and how, even – especially – at this most northerly point. With the Gulag came the Soviet state and the *sovkhoz*, the state-owned communal enterprise for herding reindeer that completed Syndassko's transition from temporary tented camp to permanent settlement with school, mayor and diesel generator.

The communist government restructured the herding way of life with quotas and fixed territories, but state support allowed it to persist as a majority way of life for whole communities until the collapse of the USSR in the 1990s made herding uneconomic. While previously more people participated in herding than in capitalist Norway, these days very few people herd any more. It is as if a tide of bureaucracy had carried the Dolgans and fixed them at the northern range of their old territory, then it had retreated and left them stranded. These days, fishing is more lucrative, much easier work and, per calorie of nutrition, far more productive. You cut a series of holes in the ice, thread the net through on poles driven into the riverbed then walk away and wait. Reindeer herding is a terrific amount of work by comparison. But it is still more high status. When the ice goes out and you can't fish any more, everyone wants to be Misha's friend. 'They all want meat!' Sergei laughs.

Misha and his father-in-law Alexei are the only two still herding reindeer in Syndassko. They are who we have arranged to visit, joining them at their winter herding camp out on the tundra.

'Can we sleep first?' I ask Dima. I have been awake for forty-seven hours.

'We are guests. We must do as we are told.' And with that three snowmobiles roar past the window and stop in a cloud of snow. Time to go.

Dima shows me how to put on the thigh-length reindeer skin boots that are the best protection against the Siberian cold. As light as cotton almost, with the reindeer's hollow fur, they are incredibly warm, like slippers, and leave no trace when you walk on the snow. Misha is wearing only a polyester skiing jacket. His wife Anna and their daughter Tania are dressed head to toe in reindeer furs with fox fur hats and mittens. Alexei, Anna's father, is driving the third snowmobile with a green canvas poncho over his reindeer furs to keep the snow off. Neither Misha nor Alexei have goggles or face coverings; they squint at the tundra like a shifty old friend.

When Dima and I are all dressed up in balaclavas, goggles and double-skinned gloves, sitting in the wooden sled ready to be towed behind Misha's machine, Sergei comes over and rubs our hats between his fingers and pinches my fingers in their gloves. Then he laughs.

'It doesn't matter anyway,' he says, 'it's not even minus thirty.'

For two hours the snow flies in our faces from the snowmobile's ribboned track, the sled bounces over the crusted surface of the tundra, and a low red sun smoulders on the horizon. We are flying. The ocean of tundra stretches out as far as the eye can see to a hazy yellow horizon, feathers of cloud against a pink-turquoise background above. The sense of space is dizzying, as if we are lost in another dimension, a white world below or above the real one.

The camp is a *baloch*, a frame tent of larch on skis made for towing behind a team of reindeer. It is half submerged beneath a snowdrift with the wooden frames of several older tents surrounding it. The *baloch* is the family's summer residence, but it has clearly not moved for several seasons.

'It's very heavy,' says Anna, apologising for the family's lack of nomadism. It needs eight reindeer to pull it. 'We just leave it here now.'

Alexei and Misha fitted it with plywood instead of the traditional larch frame with canvas and skins – it is warmer and more stable but harder to move. But that doesn't matter so much any more, since the camp has become more like a weekend cottage in the country. Anna can't remember the last time the family migrated with the seasons. When she was a girl, her father and mother lived here in the summer, and she went to school in Syndassko and, when she was older, to a boarding school in Khatanga, returning for the holidays by boat or helicopter. The family meanwhile stayed on the tundra. In the winter they would go down to the treeline at the River Papigai with many other families, but one by one the others gave up and turned to fishing instead.

At the entrance to the *baloch* is a porch with frozen reindeer skins, larch logs for firewood and frozen fish tossed on the roof. Inside, Anna lights a cast-iron stove with strips of larch. The space is about four metres by three, enough for the six of us to sleep side by side, just. Anna lays skins on the floor, puts a block of ice into the kettle to boil, takes down the radio and a car battery from a shelf and begins to sort the cutlery and odds and ends in a box by the fire. Within minutes the *baloch* is snug and warm; the ice on the inside of the Perspex double-glazed windows begins to melt and drip down the plywood interior, and I take off my outer coat, lie back on the reindeer skins and finally fall headlong into a black hole of sleep.

I wake up to see Alexei taking off his outer coat and sitting down to tea and reindeer stew by the fire. There is snow on his thigh-length boots and beads of sweat on his bald head. While Misha has gone in search of their reindeer, he has been working outside, chopping wood that he hauled here last summer, which has lain frozen all winter. He points his spoon accusingly at me.

'You say the forest is moving north? Great! We'll have plenty of firewood then.' He is not smiling. When he was young, the search for wood was a kind of tyranny. It was hard work, often involving a journey of several days. The Dolgans were not just searching for firewood; they used larch for tent poles, for sleds, for *baloch*s, for boats and oars, for tools. Almost everything was made from wood, apart from children's toys – that would have been offensive to the spirits. Toys were usually made of duck beaks or bone. Wood was special. Now of course all the toys are plastic, made in China.

'The trees were far!' remembers Alexei, who used to migrate for the winter to the Papigai with the reindeer when he was younger. It was a hard life, but Alexei is full of nostalgia for it, leading to confusion. If global warming makes life easier, he is all for it. But he doesn't want the reindeer herding way of life to end. He likes the labour-saving benefits of coal, petrol, helicopters and snow-mobiles but at the same time rejects schools, vodka, markets and smartphones. He is in some respects the archetype of a grumpy grandfather, a cynical everyman who accepts, denies and then welcomes global warming all at the same time. Is this confusion perhaps a wider symptom of the lightning pace of change in our hydrocarbon era? We have barely mourned one way of life before we are being called upon to mourn another.

Although almost no one migrates, herding is still the linchpin of the language and the culture. The Dolgans, like many cultures in transition, on the cusp of worlds, are steeped in a culture that has become uncoupled from its source. Like clinging to the branches of a tree that has been cut off at the base. Without herding there is no reason to go to the tundra, or south to the forest. There is no reason to pay attention to fine gradations of snow, to know the names for small changes in weather, vegetation, species move-ments, the signs and signals, the language and dialogue with the spirits and the natural world. For this reason, Alexei says, he will never take up fishing. But he does not blame his son-in-law for fishing – he squints hard, and the deep lines around his eyes crease in a half smile – as long as Misha herds reindeer too. Alexei was the youngest of fourteen; he is now seventy-two and the only one left. His own children have all moved away apart from Anna. She and Misha are the last inheritors of the Dolgan reindeer herding tradi-tion of Syndassko, and Alexei struggles to hide his hopes for them.

For a while Anna was seduced by 'progress'. She took a teacher training course in Norilsk, the town created for the 80,000 people that work in the world's largest nickel mine on the far side of Tai-myr. It is famous for its pollution which has killed the forest for hundreds of kilometres all around. A white dome of fumes that hangs over the city 'like a mushroom' and tastes 'like cooking gas'. It fills the mouth, 'like eating chalk', says Anna. She got headaches and hated it. The helicopter ride there was expensive. She couldn't see the point. Now she prefers not working in an office and com-ing to the clean, sparkling tundra as often as possible. It's good for her children, she says.

'When we were kids, there were no phones, our parents were in the tundra all the time ...' She tells a story of Alexei refusing to send his kids to school, the Soviet authorities hunting them on the tundra and then taking her away to Khatanga anyway. She has reverted to her father's world view, and he is clearly happy about that. But just as they have started coming to the tundra more often as a family they have started noticing changes. They have not seen huge shifts in temperature or vegetation, but they have noticed

small things, tremors of something coming. The first vibrations of
Siberia's awakening from its frozen hibernation of the Holocene
are appearing in the shape of unfamiliar sightings of birds, bugs
and butterflies and in strange bubbles under the ice.

After dinner Tania watches TikTok videos using up the remaining power on her mother's smartphone while Anna combs and
plaits her hair by the fire. She seems to be keeping her daughter
close, mindful of the seasons passing. This will be Tania's final year
in primary school. Then she will join her siblings in the boarding
school in Khatanga.

'Last spring was really strange,' she says. 'We had huge butterflies, a different species we've never seen before. The kids were
catching them.'

Daisies, swallows and dragonflies have started appearing in the
tundra, and last summer people swam in the sea for two weeks.
Usually summer only lasts a few days. The berries are getting
larger, and the sea ice in the bay in Syndassko is taking longer to
freeze in winter. And the shore by the meteorological station is collapsing into the sea! Anna says they've been noticing other things
too: ravens have started appearing, and cranes. Both birds that
they never used to see, whose normal breeding grounds are much
further south.

'What she says is true,' Alexei says. 'The seagulls are coming
earlier, then the geese and the ducks too, because the lakes are ice
free. And the little bird called *kystaatch* – the Dolgan name means
"the one who stays in winter" but it doesn't stay any more.'

Anna snatches the phone away from her daughter and shoos her
outside one last time before bed.

The sky is clear. The moonlight on the snow turns the tundra into
a glowing sea of milk. Tania is wrapped in her furs, jumping from
snowdrift to snowdrift down towards a frozen river. I wander in
the other direction towards a slight incline to take a pee in the
snow. At my chosen spot a ptarmigan, a kind of grouse with a
black tail and snowy body, shoots out from its burrow clucking
into the evening. Beyond is a flat expanse. I follow footprints to a

deep hole with a steel spike frozen in place. The hole is pure ice. I am standing on a lake. When I brush away the snow a grey marbled surface appears. Lines of fracture slicing down into blackness. This is where the water for our tea comes from. Deep down in the ice, small frozen bubbles are visible like pearls floating in the dark. Looking at the map later, in two dimensions the tundra appears as a net or a Swiss cheese, full of holes. In the summer it is 80 per cent water.

'Every tundra pond has its own name,' Alexei says when I get back inside. 'That one is called Uchukhtai in Dolgan.' This means the 'sleeping lake' or the 'dreaming lake'. He doesn't know the story, but usually there is one.

The lake is indeed sleeping. The benthic layer, the organic matter at the bottom, is held in suspended animation without decomposing because of the cold and the lack of oxygen. Frozen tundra soils are one of the largest stores of organic carbon on the planet – living matter, plants and animals that have not totally rotted away or are rotting so slowly that they have been preserved or fossilised intact. That is why Kolya and Kolya spend every summer digging for mammoth tusks in the tundra of Taimyr; melting permafrost is turning the search for prehistoric ivory into a kind of gold rush.

As the temperatures warm up and the permafrost begins to thaw, anaerobic decomposition releases methane. These days bubbles of methane have started appearing in the tundra ponds and in the sea ice out in the estuary. Those pearls are a sign that the sleeping lake is waking up.

Misha comes in with a blast of frozen mist, his lips frostbitten, his eyes pinched from the cold and a frown on his youthful face. Six hours driving in circles, but he could not find the reindeer. He seems unhappy. Alexei gives him a look and mutters something to himself.

Misha's failure could mean many things. The Cambridge-based anthropologist Piers Vitebsky has written of the neighbouring Eveny nomads, related indigenous reindeer herders occupying the tundra one time zone to the east from here, and their sense of *bayanay*, the attunement of a person to the landscape, a vast realm of shared consciousness encompassing the living world and humans in it.[11] Misha comes so rarely to the tundra, he has perhaps been unable to tune in. The next morning Alexei will launch himself and his snowmobile into the blinding snow, but he too will return empty-handed.

'Fishing is much easier!' says Anna, attempting to soothe things from her place by the fire. Misha raises his eyebrows but doesn't smile. Nothing is simple any more. Methane complicates winter fishing, he says. You have to be careful, as the bubbles make the ice weak.

'It's getting dangerous!'

In the 1970s the Kara, Laptev and East Siberian Seas were only free of ice for two months – in August and September. In 2020 the melt began in April and the sea was still not fully frozen again by the end of the year. On 18 May 2020, the *Sovcomflot*, a Russian ship, left the port of Sabetta on the other side of the Yamal peninsula from Taimyr and turned right towards China, docking at the Chinese port of Jiangtang on 10 June. It was the earliest ever journey on a northern sea route that has in recent years only been navigable from July to November.

The Siberian Shelf, which forms the seabed off the coast of Taimyr, means the ocean towards the North Pole is shallow. At the end of the last ice age, this was tundra. When the glaciers melted they inundated the land, trapping all the half-decomposed soil and vegetation beneath a cold sea that remained frozen for most of the year, resulting in methane hydrates, ice structures holding the gas. But now that the reflective blanket of sea ice has nearly gone, the dark sea floor absorbs up to 80 per cent more radiation from the sun, and the shallow water heats up fast and stays warm throughout the year. The permafrost on the seabed is melting, releasing the hydrates that get oil and gas companies so excited and make climate scientists panic. A few summers ago a drilling rig appeared in Syndassko Bay, looking to exploit the softer sea floor.

The lake brings to mind a conversation I had before coming to Siberia with Dr Ko van Huissteden, a mild-mannered and deliberate Dutch scientist, one of the world's leading authorities on permafrost. It is hard to measure methane release, he told me. Scientists have only recently been able to capture it. Sentinel, an EU satellite, can measure methane concentrations in the atmosphere, but figuring out where it is coming from is difficult. Some studies have suggested that an unstable seabed could release a methane 'burp' of 500–5000 gigatonnes, equivalent to decades of greenhouse gas emissions, contributing to an abrupt jump in temperature that humans will be powerless to arrest.[12] Ko is not sure, but that, he emphasises, is the point.

There are only four land-based monitoring stations in Siberia attempting to capture data on methane and carbon dioxide release from permafrost. And no permanent monitoring of the thawing seabed.

'You need at least ten years of data to detect anything. What's your baseline?'

Most European researchers crowd together on Svalbard, as it's hard to get to Siberia. The bureaucracy is a pain, he said.

'No one has any idea what is going on!' There is twice as much greenhouse gas – carbon dioxide, methane and nitrous oxide – stored in the permafrost as currently is in the atmosphere, enough to accelerate global warming exponentially and effectively end life on earth as we know it if it were all released at once. Yet most climate models discount permafrost because of the lack of data even though 40 per cent of permafrost is projected to be gone by the end of the century.

All this makes Ko frustrated. With governments for not investing in data: 'It's appalling how little money is going into this,' and with the media: 'The larger public still thinks that climate change will be gradual. They are not alive to the fact that it will be abrupt and what that means in terms of climate disasters and the suffering of their children.' Other scientists have referred to the frozen greenhouse gases of the permafrost as a 'monster in hiding'. Ko calls Siberia a 'sleeping bear'.

In the *baloch* large sparkling shards of the beautiful, suddenly sinister lake are poking out of the kettle, and more crystalline blocks are defrosting in a bucket hanging above the fire. While

Misha eats stew and Alexei reads his newspaper, mother and daughter play dominoes and sing quietly now that the phone is dead,
enjoying for a moment the old ways of doing things, until tomorrow morning when we fly back over the frozen tundra to that last
outpost of human civilisation, Syndassko, like astronauts returning to their space station after a walk outside in the nothingness.

Four months later, in June 2020, record heat will take the thermometer in Syndassko above thirty degrees for the first time ever.
The historic average for the month is 10–12 Celsius. Wildfires in
the region will be ten times bigger than the previous year, which
was itself a record. Collapsing permafrost will be partly to blame
for the rupturing of oil tanks in Norilsk, sending 21,000 tonnes of
diesel into the Pyasina lake and river system at the base of the Taimyr peninsula, which from space will appear red, like veins running
through the land. The *Siberian Times* will report that temperatures
in Taimyr 'broke all climate records and surprised old timers', quot-
ing an official saying that while the snow usually melts in July
there is already 'not a snowflake left in the tundra and white hares
hopping around on green ground looking bewildered'.[13] Scientists
around the world will freak out at the extreme warming in the
Arctic. And when I phone Ko in the Netherlands to get his take on
what is happening in the extraordinary summer of 2020, with
forty degrees in Europe and similar temperatures in Siberia, he will
call it a disaster and say, 'The sleeping bear is stirring.'

Ary Mas, Russia
72° 28' 07" N

This time when we return back down the river, to Novoribyne, we
find the Nganasan family at home. It has been hours, a whole day
of driving from Syndassko crawling upstream along the progressively narrowing estuary, a great sweep of white beneath a curved

sky of blue broken occasionally by a square of grey canvas *baloch* planted in the middle of the ice: Dolgan ice fishing camps. Black Kolya stops at several, seeking to trade their jerrycan of vodka for frozen fish, but they are abandoned, holes frozen over, nets and catch preserved in the frazil – almost ice – below the frozen surface of the river. It is dark before we see the trees of Novoribyne again, short spiky shadows of larch furring the edge of the world, black against the deep blue of the night sky.

We knock on another frozen wooden porch and step inside, snow pooling on the floor. An elderly woman in a velour dressing gown, slippers and very thick glasses through which she peers at us in a puzzled manner ushers us into the kitchen. She doesn't seem to understand Dima's explanation of why we are here and sends one of the many children staring at us out into the night to fetch a neighbour.

Enormous blue pipes as thick as my waist run along the inside of the wall. Pots of stew simmer gently on a solid-fuel range behind. She sits us down at a wooden table and pours out taiga tea: willowherb and mint.

'Grande Bretagne?' she shouts in halting French to me, via Dima, then in Russian. 'Is Bush your president or is it the woman? Did you see Putin in Moscow?'

She is Maria. She points to herself. While we wait for the neighbour, Maria's husband comes in.

'Yevstappi.' She points. 'Dzhasta!'

Dzhasta is tall, with powerful shoulders evident under his plaid shirt. His short white hair is closely cropped and his light eyes twinkle with the snow and ice of the tundra. Maria and Dzhasta have only lived in a wooden house in Novoribyne for the last ten years of their life. The first sixty they spent in a tent out on the tundra or in the forest of Ary Mas. Novoribyne seems to me like one of the more remote places on earth, but for Maria and Dzhasta it always represented civilisation – the place with the school, the clinic, the government office – and they avoided it as much as they possibly could.

In the old days, before World War II, most Nganasan never crossed the Khatanga River, the southern boundary of the peninsula.

Generation after generation lived and died at the top of the world, above the 72nd parallel, isolated from events on the rest of the continent, pursuing their traditional way of life largely free from the influence of tsars, Cossacks, Soviets or anyone else. They moved over the tundra at will on sleds made from larch trees in seasonal pursuit of the bounteous wild reindeer herds of Taimyr which migrated in huge annual repeating patterns up and down the peninsula – as described by the Soviet anthropologist Popov. Everything they needed was north of the gulf of the Khatanga River. And there was much to fear to the south.

Modern priorities, however, have upended the old geography. From the vantage point of governance and communication, the Gulf of Khatanga is not a border but the main navigable highway along which travelled fur traders, tax collectors and the prisoners sent to the Gulag by Stalin. When the state established itself in the two seasonal fishing villages of the nomads, Novoribyne and Syndassko, the Dolgans cooperated. The Nganasan of Taimyr did not. In the ensuing campaign against 'non-perspective' villages, many people were forcibly relocated and the old way of life effectively ended.

Dzhasta's parents were among the troublesome nomads sent to a town south of Khatanga. But they escaped and returned as close as they could get to the wild plains of Taimyr, where they had been born and raised at the time of Popov's shattering visit. The closest official settlement they could find was the Dolgan fishing village of Novoribyne.

'I was born here,' says Dzhasta, pointing emphatically at the linoleum floor of the kitchen, his voice ringing with challenge. 'On the ID that I was given, it says I was born in 1951.'

Dzhasta speaks in Nganasan to Maria, who is half Dolgan, and she translates into Dolgan. Anna, the Dolgan neighbour, has come over to translate into Russian for Dima, who then renders the words into English for me. At first Dzhasta seems reluctant to speak to us.

'I am just a wild man, a savage. I don't know much, I didn't go to school!' But at any hint of the tundra it seems his memory is unlocked.

'Before the Soviets, all of this –' he indicates the world outside the window '– our range. Taimyr!' He says the name with feeling.

'It's wide open! You can hunt!' Trees, on the other hand, were a kind of tyranny, necessary only for survival. He didn't like it. The forest was dark and closed in. They went to Ary Mas every winter because the reindeer preferred it. The forest was more sheltered; there was firewood and lichen for the animals to eat. The Nganasan followed an annual lunar cycle divided into two 'years', the summer year and the winter year, in tune with the seasonal migrations of a wide array of animals after which they named their months – 'elk month, hornless deer month, moulting goose month, gosling month' – and the cycles of the trees. *Sjesusena kiteda*, 'frosted trees month', is the second half of February and the first half of March, and *feniptidi kiteda*, 'blackening of trees month', is when the branches become free from snow in the second half of March and the first half of April, indicating that it will soon be time to leave the shelter of the woods. Then, in summer, the plateau! Fish, birds, hares, elk, reindeer. If it wasn't for a stroke he had ten years ago, he would still be out there, living in a *chum*, a tent very similar to the Sámi *laavo*. His ten children were born in a *chum*, like him. Maria nods solemnly.

Soviet colonialism after the October 1917 revolution had a different character to its Western European counterpart. The USSR did not expropriate the indigenous people of the north – it was not principally after land; it had enough of that – but sought to make them 'productive', to incorporate them into the communist economy while claiming to emancipate them from primitive customs and the feudal relationships of the past. When he returned north, Dzhasta's father joined the Novoribyne *sovkhoz*, which in this part of the Soviet empire was reindeer herding. All across Siberia reindeer herders were incorporated into the Soviet economy in 'brigades' with leaders, routes and territories based on their traditional practices and migrations but often without the flexibility and family structures of the old ways. Like other parts of the Soviet system, the brigades had targets and quotas and prizes for herders who exceeded their quotas. And the *sovkhoz* was headquartered in a town, with an office, the point of contact between the reconfigured

nomadic life and the industrial infrastructure of the twentieth century. Novoribyne was the headquarters of the brigade that
Dzhasta's father led and which he in turn led too.

In the town their family was famous. They would show up once
a year at the *sovkhoz* office, retreating for the rest of the time to
the tundra and the forest. Dzhasta, like his father, was known for
his strength. He could lift a fifty-kilo reindeer carcass by the
hooves, by himself. His mother was a fierce traditionalist, uphold-
ing the Nganasan taboo on woven clothing and speaking only
Nganasan. In Novoribyne they could not understand her. She wore
reindeer skins her whole life and was buried in the town cemetery
in 1990 with her sled and her three lead reindeer sacrificed by the
side of the grave. Dzhasta is the last Nganasan speaker in the town
and yet he is strangely matter of fact about the impending extinction of his culture.

'There are lots of holy sites in the tundra and the forest. If you
pass one you have to stop and make a sacrifice. But my kids don't
know those places because they don't migrate. No one knows
these places any more, and no one will ever know now because we
have stopped herding.'

The mythic world of the Nganasan was divided into three levels.
Larch, the 'world tree' common to all Siberian indigenous shaman-
ism, is a female deity called Mother Tree that connects the three
worlds: upper, middle and lower. The north and beneath the earth
and the thick ice is the realm of the dead, where illness and spirits
dwell. To the south is the warm home of the god of thunder. Heroes
live in the upper world, and notable points in the landscape – the
streams, trees, woods, hollows and rocks of Taimyr – are passage-
ways between the three. One should be very careful not to offend
the creatures that may live at these portals; these are the sacred
sites that no one visits any more.

The larch connected the levels. Popov's ethnology explains the
centrality of the larch tree:

> On the seventh day Dyukhade reached the highest level of the sky.
> A long pole was erected in the middle of the tent. The shaman

climbed up and poked his head out of the smoke-vent. The pole symbolised the tree, rising in the centre. On the top of the tree a spotty-faced deity was living ... Then he was carried to the shores of the nine lakes. In the middle of one lake was an island. On the island was a tree, a larch rising to the top of the sky. It was the tree of the Mistress of the earth ... Then he heard a voice: 'It has been decided that you shall have a drum from the branches of this tree.'

The job of the shaman was to mediate between the levels and ensure that balance and respect were maintained. They would commune with the animate world that sustained all life using the drum and song. Every living thing, including humans, had its own unique autobiographical song essential for calling it into being or addressing it in rituals. Dzhasta remembers shamans in Ary Mas, their houses apart from the rest of the tents, made of larch logs and earth. As a child he was instructed not to shout or make a noise near them, to pass at an angle and never look directly at the shamans and their dwellings.

'Did you know Kosterkin?' I ask. Kosterkin was a famous shaman, the last shaman of the Nganasan, who died in the 1980s. He once allowed his week-long seances and teachings of Nganasan animist beliefs to be filmed by an Estonian TV company.[14] This is one of the only records of traditional Nganasan shamanic practice. The mythic world of the Nganasan was very sophisticated. There was no distinction between spirits and the physical world, and the animating life within each plant, rock, person or animal was a spirit not tethered to the physical body of the thing concerned. Each set of spirits had a code of eight rules, customs, language and clothing. Beyond this, smell was critical. To be exposed to the smell of something for too long risked absorbing the qualities of that thing. So a human sitting next to a tree for a long period could make that tree become human. But Dzhasta gets cross when I mention the shaman.

'I heard of Kosterkin but I never met him. All that stuff is gone now. The Russians!'

When I ask about traditional medicine I get the same dismissive reaction.

'We had the best vets and doctors; the Soviets brought them to the tundra in helicopters!'

There seems to be a kind of war taking place within the bodies of those elders who have early memories of a way of life unchanged since the ice age yet whose adult experience has been one of lightning progress that has made their lives easier even as it has destroyed so much. They have been colonised by hydrocarbon thinking and must accept the Faustian bargain and the circumstances of their descendants, painful as it might be. Alexei wears his confusion on his sleeve, flipping between longing for what has been lost and acceptance of the cost. But Dzhasta is more rigid, clinging to the story of progress as retailed by the Soviet state despite what it means for his heritage.

When I ask about his thirteen grandchildren and three great-grandchildren, who will never speak the Nganasan language, Dzhasta shrugs his shoulders, as if to say, 'So what?' Through the door to the bedroom, they are just visible, the small faces of the descendants of the proud foresters of Ary Mas lit by the blue glow of a television gurgling away in Russian.

Dzhasta's attitude recalls a study of Nganasan culture I read in which Professor Helimski of Hamburg refers to the 'Iced Culture of the Nganasan' as a relic preserved in the permafrost, an elaborate oral culture developed over the many long months of winter that prized storytelling, grammatical precision and metaphorical skill. Helimski quotes elders reluctant to fight against the tide of change, saying of their children, 'Let them better not speak our language at all than butcher it.'[15]

At first glance it seems a strange position, but there is something in the proud refusal to compromise, a nobility but also a different relationship to mortality. For indigenous people living close to the vast and unpredictable forces of wild nature, death is ever present. Perhaps there is a freedom in acceptance that assists the mind in travelling beyond the narrow window of a single lifespan or a single species, which allows one's self to be fully integrated with a magnificent, all-encompassing whole: we are nothing, but we are everything. It is an awesome, challenging prospect. Dzhasta's sense of time comes from a similar perspective. It is geological, a sensibility that the rest of us might soon find helpful.

'The trees are coming north, you say? Well, do the scientists say there were forests here before?'

They do, although there is dispute about the time frame. To correlate Dzhasta's oral history with the geological record would be the real scientific prize, although I suspect the prospect of such a conversation is remote. Humans are a blip in the record. And Dzhasta is not nostalgic about them. His acceptance of nature's harsh regime is total, and it leads to a humility that is courageous in its expectation of challenge and uncompromising in its rejection of sentiment.

'Global warming? Tell my grandchildren, I'll be dead by then.'

With the interview over, Kolya and Kolya decide that if we want to see Ary Mas in the daylight, we must set off 'early'.

'Like 4 a.m.? Three?' I ask hopefully.

'No, midnight.'

So, after another meal of iced fish, mustard and chilli powder with Anna and Konstantin, we set off through the night following a different frozen river, the Novaya, a tributary of the Khatanga, north into the heart of the peninsula. I see nothing, am aware of nothing except the bruises on my body as the Trekol tosses us around like peas in a rattle.

We arrive at moon-set, nine hours later. The pale lunar disc floating in an indigo sky is scarcely different in hue from the expanse of snow-covered tundra below. Darker flecks of grass and the occasional lone tree mark the ground, giving a sense of scale. Without them, the undulating white waves are disorienting, stretching in every direction like a white Sahara Desert. We are looking for a forest in this huge, endless landscape. Where are the trees? They should be here, says Kolya.

Kolya is staring at a little GPS device held together with tape in one hand while the other grips the wheel of our strange truck.

The moon casts an eerie glow as the truck whines on, following the course of the river, and we peer out of the misty windscreen at the folds and ridges of the snow piled along the bank. Then, suddenly, the forest looms into view, surging out of the tundra, flooding the dips and troughs of the valley below. This is odd, for although we are driving on a frozen river, we are above the forest. But very little seems to make sense this far north, just before dawn, on 11 February 2020, at forty-four degrees Celsius below. I have spent the night rolling back and forth in the pit of the truck, crashing into diesel tanks, toolkits, spare parts and a five-litre jerrycan of vodka without sleeping a wink. But now I am wide awake, smiling at the dawn. The rigours of the journey are forgotten. We are finally here.

The trees fall out of sight for a while; then, around another bend in the river, they rise up a ridge. Rank upon rank of spindly stems tower above us, backlit against the yellowing haze of the gathering dawn. Only larch. The thin branches and delicate needles give the frosted *Larix gmelinii* a frail aspect, but this is misleading. In fact they are the toughest trees of all in the taiga, and therefore the world. The only species that has evolved to survive this far north, in this extreme cold, where the permafrost is over 200 metres thick, and the temperature has always stayed below freezing for nine months of the year.

I jump up to the front, excited to see the forest emerging from the pale blue pre-dawn light. I have read and dreamed about this climax of the treeline for months. Russian pop music continues to burble away, and the swaying of the Trekol now feels almost triumphal as we approach our destination. Blond Kolya could not understand the point of visiting the forest and tried to dissuade me several times from this detour. But now he is smiling too.

The forest continues to our left, keeping us company on the rising southerly bank of the river. We drive on, west, until more trees appear on the northern bank along with a collection of snow-blasted huts, a washing line, a weather station with an aerial and a frosted metal sign driven into the ground reading, in Russian, ARY MAS: THE MOST NORTHERLY FOREST IN THE WORLD.

Blond Kolya reaches down for the spanner that serves as a gear stick and pushes the truck into neutral but leaves the engine running. It has been running continuously for four days. He turns off the music that has been squeaking from his phone all night, sits back in his seat and exhales loudly. He has not complained, but the journey across the trackless, frozen tundra with only one poor headlamp and very little sleep has been a trial even for someone as experienced as him. He puts his forehead on the steering wheel inches from the dashboard with its shallow basin of keys, electrical tape, safety pins, matches, lighters, a USB stick, five bullets and a bottle opener and reaches into the driver's door pocket for a battered pack of cigarettes. He lights one and opens the door, ushering in an icy blast, and turns to me in the passenger seat with his playful half-smile.

'Ary Mas! OK? OK Ary Mas!' He doesn't speak English and I don't speak Russian, but we understand each other perfectly. What he means is, 'Are you happy now?'

We pull on our snow boots, hats, two pairs of gloves, balaclavas and insulated jackets and get out of the truck. This is a complicated procedure. One must hold the door while stepping first onto one of the huge spherical rubber wheels of the Trekol, then jump a metre and a half down onto the frozen surface of the river.

The riverbank rises steeply towards the scientific station. The snow is turning a creamy orange colour to match the lightening sky, and out of it poke crusted and crystalline grasses and seed heads suspended in the delicate air. The frozen river arcs away to the west, leaving this little island of trees on the north bank marooned, all by themselves. An island within the island of forest. Some 15,611 hectares of trees are surrounded by tundra on all sides. No one knows why the forest has survived here. The most popular theory is that the last ice age was particularly aggressive, locking up large amounts of earth's water in ice, lowering sea levels so much that the shore of the Arctic Ocean was hundreds of kilometres to the north. This left the flatlands of the Taimyr peninsula unevenly glaciated, with Ary Mas a relict from the previous ice age.

Others believe that the trees are relative newcomers, taking advantage of favourable soil structures, and are very slowly on their way further north still. But they grow and die so slowly, neither theory is easy to prove on human timescales.

Dima and I crunch through the sparkling surface of the snow. The huts are deserted; no one has been here all winter. The trees are stock still, with not a whisper of wind. The washing line hangs slack. The moon is down below the horizon now and the sun is just above. The snow is glistening pink, and the needles of the trees seem on fire. Larch sheds its needles in winter, but clearly the first frost came before the trees were ready, and the dead, desiccated needles are frozen in place. I brush one of the larch, and the needles tinkle to the ground at my touch. The slender branches are covered in very fine trachoma hairs, similar to those of the downy birch, a fur coat to trap heat and modulate the cold. They snap like balsa wood as there is no moisture inside the twigs whatsoever. This is why larch makes such good firewood in winter, and why they can survive in these temperatures at all: they have evolved a mechanism for avoiding the fatal formation of ice crystals inside the living cells of the tree.

The foremost larch specialist of Ary Mas, Nadezhda's friend Aleksandr Bondarev, was jealous that we were visiting the forest in the winter. He, like most other scientists, only ever does research in the summer. 'My trees! They will remember me! Greet them!' he had urged me on the phone. His enthusiasm for the Dahurian larch is uncommon, born of twenty years of getting to know them, on and off. 'It is a very clever tree. Smarter than you!'

Larix gmelinii is a wonder of the natural world. As winter approaches, the tree begins to draw water from the xylem, the capillaries in its trunk, into the bark and all the other spaces outside the living cells of the tree. This allows the plasma membrane of the cell to deflate and ice to form outside the cell, without harming it. As the air temperature drops to minus five or six, the temperature within the stem remains at zero as the tree cools slowly. The temperature gradient between the liquid within the cells and the ice outside them allows vitrification – the solidification of water without it turning to ice. This is what happens if you put a tree in liquid nitrogen: super-cooling turns the water to something akin to glass rather than ice. Ice is made of crystals which cut and slice up the cells, leading to a condition called freezing injury. Vitrified water on the hand is smooth and solid. This slow cooling allows the water to be drawn out of the living cells. Meanwhile many other changes take place. The tree floods its cells with abscisic acid, which increases the permeability of the cell membranes to water, allowing them to 'leak'. Sugars and proteins change at low temperatures into depolymerised sugars that bond to the frozen water so the tree can resist dehydration. As it gets colder, the larch produces chloroplast lipids to lay down fat and enhance the plasticity of its cell walls, further reducing the water content. There is so little moisture inside a hibernating larch tree in the middle of winter that it is impossible to tell if it is dead or alive.

It is this ability to manage moisture in the form of water and ice that allows the larch to thrive in the comparative desert of Siberia. In the winter their roots may be frozen solid, but even a short day of winter light on the tips of their branches can activate the roots, drawing moisture up into the branches and needles from the iron-hard ground. At low temperatures moisture is scarce, but ice – in the form of permafrost – is abundant. Dahurian larch likes the cold so much that researchers believe it co-evolved with the spread of permafrost. It is no surprise that this cryolithic forest, the frozen forest of the northern taiga, should be dominated by the species that has learned to love the ice the best.

Beyond the sparse buildings of the science station, where blond Kolya is examining claw marks on a smashed-in door, apparently the work of a wolverine, the trees grow stunted, thin out, and stop amid knee-deep snow. I stare away from the rising sun out towards the tundra emerging from the dawn, the beautiful, unforgettable, crisp edge of the forested world. It is an abrupt treeline. In the long dawn shadows the trees appear as arrested figures wandering in clumps of twos and threes and finally several solitary outliers a few hundred metres out, alone in the tundra, stumbling towards the pole. Then, as the sun inches above the canopy of the forest, it appears to torch the short branches of the wispy larches, and suddenly there are flames of orange and red flickering over the mauve snow against the softest pink-blue sky.

The pattern of trees revealed in the early morning light appears to show pioneers of the taiga striking out towards the Arctic Ocean, sending scouts ahead to survey the terrain, but talking to Aleksandr has given me new eyes with which to view this part of the treeline and the forest of Ary Mas.

Aleksandr began studying the trees of Ary Mas in 1995, but after five or six years was sent to another part of Siberia to lead conservation efforts in the Altai Mountains. He returned in 2019 after a break of nearly twenty years to find his beloved larches had neither moved nor grown.

'It was very strange for me,' said Aleksandr.

The maximum increase in diameter was two millimetres, and for most of the trees that he drilled and took core samples from to count the rings, not even that. The trees were the same height and the view exactly the same as it had been all those years ago.

Aleksandr's findings contradicted all the models of treeline advance, upsetting the image of forests leaping north with which I began my research and which simplistic models based on temperature alone predict. When a Cambridge-led study collated all the research on the treeline over the last twenty years, the results, published in 2020,[16] showed a mixed picture of species responding differently in different ecosystems. And Aleksandr's observations chimed with two other reports – from western Siberia and eastern Alaska – where the treeline appeared to be stable or even in retreat. Trees are not simple machines that photosynthesise more carbon dioxide and grow more just because it is warmer. The larch's adaptive DNA might well be conditioning another response. Survival in this extreme environment is not determined by growth or size or seed load, but strategic thinking. It is, as Aleksandr says, a very clever tree.

There is as yet no appreciable sign of warming in the tightly packed tree rings of the larch of Ary Mas. Winters have been getting warmer, and average summer temperatures have been creeping up slowly, but the winds that blow off the sea ice have, until recently, always kept summer temperatures down. But even if the absence of treeline advance could be explained by other factors such as the poor quality of the soil, the lack of available nutrients or an insufficient mycorrhizal fungal population to supply an increase in the number of trees, Aleksandr had expected more filling out, more succession within the forest. But this was not the case. The understorey was just the same as twenty years earlier. And there was very little dead wood. He put that down to Dzhasta's nomadic herding brigade, which was only banned from entering and cutting the forest once it was made into a national reserve in 1979.

'They were excellent foresters!'

But still the understorey was uncommonly sparse.

'It was puzzling. Ary Mas is a very interesting forest,' he said.

The forest here is open. If it weren't for the snow that drags at my knees, I could stroll easily through the short, krummholz trees. It is not a closed-canopy wood in the traditional sense – the branches of the trees are not touching. But it is a closed canopy in another sense: underground. The distance between the trees is determined by their roots, spread out in the very shallow active layer of soil above the permafrost, the thirty centimetres which thaws for a hundred days or so each summer. There is only so much soil to go around, and the larch appears to prevent other shrubs from getting a toehold between the trees. Investigating this strategy, Aleksandr made another discovery of the unique nature of Ary Mas.

Further south in the taiga, other species of larch have relatively limited root systems and more phytomas (living cells) above ground. But in Ary Mas, almost half of the plant's phytomas is below ground and half is above. The ground covered with snow is insulating in winter and protective, out of the wind, so more of the tree is underground. It is also closer to the permafrost, the source of nutrients and moisture. Excavating the neighbourly relationships of this strange underground tree, Aleksandr noticed that beneath the mossy layer of un-decomposed needles and lichen, the larches of Ary Mas were reproducing vegetatively – without pollination.

The roots are suckering and new shoots above ground are connected to the roots and stumps of existing trees. All the trees appear to be working in concert.

Moreover, Aleksandr noted that all the mature trees seem to have a shared height limit of five metres, regardless of age; there are no specimens higher than that. In larch forests further south, mixed heights and stands are the norm. If you take a bird's-eye view of a larch tree, it has a corkscrew character, the branches radiating out in a spiral as they go down the trunk. This allows the branches to catch as much light as possible without shading others. We don't know for sure, but it seems that in Ary Mas the evenness of height and space between the trees allows the maximum amount of light even at low angles to penetrate the forest and reach every tree. As the tips of the trees catch shards of pink dawn on their iced needles, the patterning of the treescape suggests a collective intelligence at work. The forest here took very long to form, adapting in unique ways, evolving a system tailored to this extreme environment. It seems entirely reasonable that this intelligent distributed organism should be cautious about making any sudden moves.

Temperature anomalies for Siberia in 2020 are frightening: at four times higher than the global average they are the highest in the world. Above the 75th parallel, in 2020 the Arctic as a whole warmed even faster, at six times the global average.[17] But the warming is starting from such a low base it can be hard to notice. Today it is minus forty-four, just inside the normal range for February of minus forty to sixty, but, as the Dolgans and others who live outside every day know well, temperatures below minus forty are rare these days. Large masses of land and water take a very long time to heat up. While the leading edge of climate breakdown is being felt first in maritime areas with a high sensitivity to fluctuating weather patterns like Norway, or, as we shall see next, Alaska, in the deep freeze of Siberia everything happens far more slowly.

And so it is easy to believe, as Dima my translator does, that global warming is a hoax intended to economically cripple hydrocarbon-rich Russia, where, the saying goes, 'Oil is our father and gas is our mother!'

'Greta [Thunberg], she's just a puppet, she's being coached, right?'

There is nothing in the pristine view this morning to suggest alarm. The dawn is utterly silent. The delicate crusted surface of the snow is covered in a fine lacework of tracks. The wind here wipes the stage clean each night; what is written on the snow is the fresh drama of a single morning. An Arctic hare has been digging for frozen berries, splashing the snow with what look like flecks of scarlet blood. It loped off in the direction of the river as other animals appeared: several Arctic fox, ptarmigan and a single wolf. The wolf prints are huge and steady, long toes almost the size of my palm compacting the snow about an inch deep. She was not in a rush. A whole society of forest life with an ensemble cast of creatures was here but moments earlier. They must be close by, hidden in the trees, watching. The stubby ranks of larch are not passive. It is a forest of eyes. I feel like an intruder in a private world; the pink snow, the motionless trees, the pitiful human-built shacks seem incidental.

'The taiga is mighty and invincible,' wrote Chekhov. 'The phrase, "Man is the ruler of nature," nowhere sounds so diffident and false as here.'[18]

When the Trekol heads downstream a few hours later, the sky is turning orange with the end of this short winter day. Approaching Khatanga again, the streaming plumes of smoke from the power station are illuminated high in the atmosphere, their gold and violet stripes indicating the height of the temperature inversion, painted lines of sky half a mile up. At the first sight of this familiar pollution, blond Kolya gets excited, opening the window and lighting a cigarette.

'Welcome to my beautiful modern city!' he says, spluttering with laughter.

Dima is triumphant that the trees of Ary Mas are not on the move, as though proving that my claims of changing climate are indeed a conspiracy.

'You see!'

I have tired of trying to convince him and so I just smile instead. The next day, however, when we pay our final visit of the trip to Bondarev's friend at the offices of Taimyr National Park in Kha-tanga, Dima is uncommonly quiet. The trees might be slow, but there are other signals of warming, other species that can move faster, that have legs or wings.

In a poorly heated wooden office above a room full of natural history exhibits, we meet Anatoly Gavrilov, the last remaining staff scientist of Taimyr National Park after the government 'optimised' costs and reduced the number of scientists from sixty-seven to thir-teen, then eleven, then one. The choice of Anatoly is telling. He is an ornithologist.

Taimyr, as the most northerly continental land mass, is at the apex of five of the planet's eight major flyways – migration routes for birds. Birds that winter in Australia, west and southern Africa, the UK, the Mediterranean, India, China and Central Asia, all come to the top of the world to breed. The boreal is home to five billion birds, and Taimyr has the highest concentration of bird spe-cies in the world. Every summer Anatoly spends sixty days based at the wooden huts we had visited at Ary Mas (and where he had got to know Bondarev), scouring the forest and the tundra for birds. Every year he records dozens of new species that he has never seen before. On his desk is the feather of a raven, a new-comer to the peninsula. But that is old news already. In the last few years he has had many more surprising sightings than that. More and more southerly species are appearing in the forest and the tundra north of Ary Mas.

'The forest is getting crowded!' And new kinds of grass species are bringing new insects, which are in turn bringing new birds. Tropical species from China, for example, new kinds of seagulls from the Atlantic. The most shocking species he saw was a hoo-poe, which belongs in the temperate forests of the Mediterranean and Black Sea.

'I could not believe it! Never would I have expected that.'

The trees may be stationary, but the rest of the forest is shifting north. Anatoly is still pulling out maps and searching for English translations of species when Dima excuses us – our plane is leaving soon. Anatoly seems genuinely sad to see us go, as though he doesn't get many professional visitors any more. There are few audiences for his message: the birds are warning us.

'I am a lone voice in the desert up here! In Russia people have this strange idea that their actions change little in the world.'

Dima translates this last sentence then looks away sheepishly. At the airport he wants to know: 'So how bad is it?' I explain that we don't really know.

'Where should I move to?'

'The same direction the birds are going – north.'

The plane is delayed, and we are trapped in the freezing tin box of the airport building while a dog with shit frozen to its backside circles us endlessly. I feel nauseous, but not just because of the smell. Airports are like the Nganasan portals between worlds, where the contingency and threat of our fragile lives is evident: many branching futures are possible. Dmitry's question and my answer bring the weight of the facts we have witnessed into personal focus. This is not a process that we are observing from afar. When we get home, we will not be safe. I feel a lightness in my limbs and the constriction in my chest familiar from working in war zones. It was how I felt when the rockets whistled overhead or the shooting got too close, when the hand of the militia man at the checkpoint went up. My instincts are telling me to run, but where?

Cherskiy, Russia
68° 44' 23" N

In the summer of 2018 an unconventional earth scientist with a beret, beard and cigarette called Sergei Zimov, looking more Left Bank philosopher than physicist, sent shockwaves around the scientific world. *National Geographic* magazine featured photos of a grinning Sergei at the controls of a pneumatic drill, boring through

the permafrost at his research station in Cherskiy in the far north-
eastern corner of Siberia.[19] Cherskiy is on the treeline, in the zone
of continuous permafrost, four time zones east of Taimyr. It is
where the famous gold-bearing Kolyma River, one of the bastions
of the Gulag and the most easterly of Siberia's great rivers, meets
the sea in a wide flat delta of tundra, permafrost and light tree
cover. On the same longitude as Kamchatka and New Zealand,
another one time zone east will bring you to the Bering Strait.

It was not the drill itself that caused the shock – Sergei is known
for his unique research methods – but what he found. After a metre
or so of solid ground, the soil underneath was slush, just as he
predicted: the permafrost was melting from below as well as from
above. Almost everyone had assumed that the permafrost would
gradually melt from above, as a result of warmer air temperatures
deepening the active layer of unfrozen soil on top. But Sergei's
finding pointed to a different future: rapid permafrost collapse. It
meant any frozen ecosystem based on permafrost was in urgent
trouble, not to mention the unquantifiable emissions of methane
and carbon.

Dahurian larch does not like getting its toes wet. It sends out its
roots horizontally into the shallow soil on top of the permafrost,
drawing the moisture from below. It will tolerate water for short
periods, when the active layer, the top 30–100 centimetres, melts
for eighty days or so in the summer, but the larch's root activity
requires oxygen, and prolonged waterlogging spells death for the
tree. Ice, on the other hand, thanks to the air pockets that freezing
opens up within the soil structure, brings life. In the deep freeze of
Taimyr, the permafrost is intact and the larch is stable, but in far
eastern Siberia, where the warming currents of the Pacific enter the
Arctic Basin and bring warmer patterns of circulation, the treeline
that runs all the way from Taimyr past the Laptev Sea, across the
vast watersheds of the Lena, Yana, Kolyma and Indigirka, is actually in retreat.

In some cases where the permafrost melts, the water drains
away and the land collapses, leaving great sinkholes, a common
image now across Siberia with the larger ones looking like craters

made by giant meteorites. In low-lying areas, however, the water has nowhere to go and the water table rises. As the scientists of the Sukachev Institute found, a water table of 1.5 metres or higher is fatal to larch. The waterlogging in the soil 'drowns' the trees. Without the permafrost, the larch is vulnerable and outcompeted. Along the banks of the Kolyma River where the permafrost thaw is most marked, willows, poplar and birch are gaining ground.

Sergei Zimov's son Nikita runs the North East Science Station now. Nikita had promised to take me out in a boat later in the summer to see the melt first hand, along with other scientists measuring methane and carbon emissions from permafrost as well as the amount of organic carbon being carried by the rivers out to sea. The Zimovs' facility is one of the four mentioned by Dutch permafrost expert Dr Ko that are trying to quantify methane release. But Russia has now suspended all flights due to the Covid pandemic, and in any case the airport in Cherskiy is out of action – forest fires have blanketed the region in smoke. So when I finally meet Nikita it is on a screen.

'Larch? I hate larch! You called the wrong guy,' says Nikita, laughing. 'It is very good for the stoves.'

Nikita and Sergei don't give a damn about saving the forest; it's the permafrost they are concerned about. And strangely the best way to slow the thaw and perhaps preserve some parts of the taiga is, it seems, to cut the trees down.

As he talks, Nikita spins in a large leather-backed chair in his wood-panelled office in the research facility Sergei and his colleagues built in the 1970s. On the wall behind him is the skull of a bison with horns intact, while leaning against the doorframe is an entire mammoth tusk, taller than him. He is wearing a 'Florida State' baseball cap atop a round bowl of mousy brown hair and a red Nike T-shirt. It is June 2020. Siberian wildfires are making headlines around the world, and Nikita's videos taken from a nearby hill show fires raging across the horizon, sending a continuous band of smoke over the town and blotting out the twenty-four-hour summer Arctic sun. Nikita is locked down alone at the research station. His wife and daughters are 'on the main-land' in Novosibirsk in southern Siberia. His father is near Moscow. But someone must look after the atmospheric measurements and long-running experiments; the research station is their family busi-ness in post-Soviet Russia. Normally at this time of year there are dozens of visiting scientists.

Nikita was born in 1982 and grew up on the research station in the company of scientists from all over the world. His childhood winters were colder, and the ice in the estuary always broke up and went out to sea between 1 and 10 June. 2020 was a new record: 24 May. During the Soviet era scientists were an elite; postings to Cherskiy were coveted and well rewarded, and the school in the town was good. But after perestroika and the collapse of the USSR, things went downhill, and Nikita completed his education on the station being taught chemistry by the wife of one of his father's colleagues. He won a scholarship to university in Novosibirsk and studied mathematics but didn't like it. 'There are twenty-seven fields of mathematics at the university, and twenty-six is already too much.'

He found himself using mathematical models to look at ecosys-tems and environmental change, and his father persuaded him back to the family business. Twenty years later, he is still here, continuing Sergei's vision and scientific agenda, though he doubts his daughters will want to keep the tradition alive.

Sergei Zimov set out from the Far East University in Vladivostok to establish the North East Science Station at Cherskiy in 1977. He went north, Nikita says, to be in nature, to escape Soviet bureau-cracy and to indulge his passion for hunting. 'Except by the time he realised Cherskiy was not a great place to hunt, it was too late.'

In a 2006 article in *Science*, Sergei Zimov was one of the first scientists in the world to warn about the amount of carbon and methane stored in the frozen Yedoma soils of Siberia and the sen-sitivity of those soils to warming.[20] Yedoma is permafrost that contains partially decomposed soil: the leaf litter of the forest that has rotted very slowly or not at all due to the lack of warmth and fungi to break it down. Yedoma contains five times more carbon than the floor of the rainforest. And permafrost covers far more territory than the rainforest. Until it started degrading, permafrost underlay one quarter of the land surface of the planet, with over half of that found in Siberia, and it is melting much faster than models predicted.[21] The importance of trees here is not only the oxygen they produce and the emissions they sequester but the role they play in slowing or accelerating the melting of permafrost.

Sergei made a name for himself with bold practical experiments such as clearing the active layer off an acre of permafrost with a bulldozer to see how it would degrade. A decade later, that ground today is a pit containing huge pillars of mud – all that remains when the water and soil melts away. And he advanced some radical ideas about the larch of the boreal forest which were initially considered controversial but are now mainstream.

Building on Sukachev's insights from nearly one hundred years ago, Sergei suggested that twenty million years ago there was a revolution in ecology. Tall plants – the ancestors of today's trees – learned to defend themselves from browsing by producing poisons. Grasses meanwhile formed an alliance with herbivores. All herbivores depend on grasses, which evolved to depend on herbivores for fertiliser and reproduction – the dispersal and germination of their seeds. This triangular ecosystem – forest, grass and grazing animals – was productive, and durable for millions of years. Geological evidence suggests there were ten times fewer trees in Siberia 15,000 years ago. As the African experience of the Serengeti and other European experiments in biodiversity such as those at Oostvaarderplassen in the Netherlands and Knepp Estate in the UK seem to show, a closed-canopy forest is often not the climax ecosystem, rather the browsing of megafauna – elephants, mammoths, elk, moose and aurochs – created a mosaic of mixed woodland and grassland. Zimov proposed that the taiga had in fact been a savannah. Then a super-predator called *Homo sapiens* emerged and took out too many of the herbivores in the chain – the mammoths, bison, elk, horses and so on – and without them the grass was outcompeted by shrubs and trees, above all by the larch, which had evolved so successfully to thrive on permafrost. It is an interesting theory: that the apparently timeless larch taiga is in fact a geological upstart – a weed unleashed by human activity.

'People see the taiga as a wild ecosystem, but before the first humans came, trees were very rare in this place. Larch is a useless ecosystem. No one eats larch; only rodents eat the seeds,' says Nikita.

Subsequent research has confirmed Sukachev's insights about the youth of the taiga and Sergei's larger theory about what had gone before. Since the end of the last interglacial, the cryolithic larch forest has evolved in dialogue with the permafrost, putting out its horizontal roots in the active layer above the frozen mass below, creeping north to form a single-species forest from Siberia's border with China 2000 miles to the south all the way to the northern treeline. This is a dangerous, fragile and man-made situation, according to Sergei and Nikita.

With larch on top of it, the permafrost is much more vulnerable to warming because larch traps snow. Nikita maintains daily temperature records at the station, continuing the work that Sergei started when he was a child. In Nikita's lifetime, the temperature of the permafrost has risen from minus six to minus three degrees Celsius. At the same time, the average air temperature has risen from minus eleven to minus eight.

'That three-degree warming of the soil is caused by snow,' Nikita explains. The winter cold does not penetrate into the soil because of snow cover.

Sergei believed the lost animals of the savannah would have trampled the snow to reach the grass or brushed it away entirely, considerably reducing its insulating properties and allowing the cold air to refrigerate the soil. Wind, as I saw in Finnmark, Norway, would do the same – if there were no trees. Sergei posited that permafrost under forest was several degrees warmer than permafrost under savannah; he just needed a way to test it. So, in 1988 he acquired sixty-two square miles of scrubby forest-tundra treeline from the Russian government to establish Pleistocene Park, an experimental safari park just outside Cherskiy.

The park is stocked with six species of large herbivore (horses, moose, reindeer, musk ox, elk and bison) to replicate the grazing activities of Pleistocene savannah herds and to demonstrate that converting the taiga back to tundra-steppe is the best way of slowing permafrost thaw, thus buying humanity a little more time to avert catastrophic global warming. As expected, the animals crushed shrubs, moss and seedlings and encouraged more grass, and data showed that the soil of the park was indeed up to two degrees colder than the forest. And two degrees can make all the difference.

In 2018, the year of the pneumatic drill experiment, the active layer did not refreeze at all. In 2019 it re-established but it was a close call.

'This has deep implications in our region. There will be massive, sudden, permafrost degradation soon, which will be, how shall we say? Not so great,' says Nikita with a chuckle.

All the towns, roads and pipelines of Siberia are built on permafrost, and they are already collapsing. Life in Siberia as currently built will be impossible; a whole new infrastructure will be required. Mud slides are carrying homes and tracts of land into the rivers. All this melting soil in rivers is changing the hydrology, affecting aquatic life. Fish have stopped migrating; the native white fish have all but disappeared from the Kolyma, and seals have appeared in huge numbers. And it is not just the life of the rivers that is changing. Siberia's rivers are discharging 15 per cent more water into the ocean than a decade ago, and that is increasing year on year. This seems to be changing the salinity of the Arctic Ocean, and may in turn affect the Arctic pump. This is the process by which salty water sinks to the bottom, causing a cycle in which the deeper water mixes with nutrients from the sea floor and rises again to the surface to feed phytoplankton. This process of primary production, in which nutrients are turned into plant and animal matter, is the bottom of the oceanic food chain. This stimulation of plankton growth is also the reason the entry points to the Arctic Ocean – the Bering Strait and the Barents Sea – are among the richest feeding grounds on earth for marine animals and birds.

Nikita has also inherited his father's bitter take on the conclusions their research has yielded. 'From a personal perspective I will not cry about the end of modern life, but it will be a shame if the fish are gone.'

The research station is built on a former quarry, a tribute to his father's foresight. Nikita will be fine.

'But if the local town slides past my window it will be sad.'

Sergei projects the collapse of human civilisation soon, because he cannot see how the warming will stop.

'But I don't think he's frustrated by that,' says Nikita. 'He hopes for it as soon as possible, in order to prove his hypothesis. He is a scientist!'

But contrary to their apparent misanthropy, the Zimovs are not cynical, they have not given up. Nikita is planning on chartering a ship later in the year with a French TV crew and heading to the remote wildlife haven of Wrangel Island in the Bering Strait to collect musk ox for Pleistocene Park. And Sergei is in Moscow, where he is setting up another visitor attraction, called Wild Field, to try to educate people and change attitudes. Despite the Zimovs' scientific success, the example of the park has not yet sparked large-scale emergency schemes to replicate its cooling effects. Nikita is not surprised: 'If it's hard in Russia, where the government owns everything, it must be very complicated in other countries!'

And cutting down millions of acres of forest to replace it with grass is such a counter-intuitive solution to global warming it is unlikely to take hold in a world trying to arrest deforestation and plant more trees.

Still, we desperately need the challenge of such wild ideas to grasp the peril and the possibilities of the present moment: we helped make this, and that means we can unmake it or make it anew. What the Zimovs' work shows is that humans have been as much the keystone species driving the ecological succession of the taiga as the larch – if not more so. As with Scotland or the memories of the Nganasan, we must learn to look and see with ancient eyes. The landscapes we have grown up in and taken for granted for a few short generations are not timeless at all, but a human-shaped moment in a continuous dynamic of changing colours of blue ocean, white ice and green forest on a ball of rock, surrounded by gas, spinning in space.

4. The Frontier

White spruce, *Picea glauca*   Black spruce, *Picea mariana*

Fairbanks, Alaska
64° 50' 37" N

Just fifty miles of shallow water separate Russia from Alaska. During the last glacial maximum, 20,000 years ago, with much of the earth's water locked up in glaciers, sea levels were a hundred metres lower, and the Bering Strait was dry land. In geology, Alaska and the Yukon are known as Eastern Beringia – essentially considered extensions of eastern Siberia. When the land bridge was intact, plants, animals and humans travelled freely across a tundra-steppe ecosystem dominated by grasses and sagebrush with scattered groves of poplar, and, as the ice came down across Eurasia, hundreds of species including early humans moved across these lowlands from west to east, into the cul-de-sac of Alaska and the Yukon. Glaciers in the Brooks Range, the mountains to the south of Alaska and the Canadian Rockies, stopped further movement inland, making Alaska and the Yukon today one of the most biodiverse areas of the Arctic tundra belt with over 600 species compared to Taimyr's 118.

When things warmed up and the sea cut the land bridge joining the Pacific and the Arctic Oceans, the ecological history of the planet was changed for ever. Dozens of species found themselves on either side of the Bering Strait. Like twins these two separate ecosystems shared many common characteristics, but in a warming climate it only takes a difference of one or two species to cause systems to evolve in strikingly divergent directions. Alaska has two keystone species that north-eastern Siberia does not: spruce and beavers.

Alaska seems to prove the Zimovs right. Their theory that the larch forest of the taiga is a young geological formation, probably in response to humans taking out the megafauna, is supported by the absence of larch on the other side of the water, which must have come from further south once the sea had already broken through the Bering Strait. The forest on the Alaskan side has no larch at all, but is dominated by spruce (*Picea glauca/mariana*), whose origin too it seems was further south, suggesting a connection to the interior.

Spruce are hardy conifers with tough waxy needles, tall spires with short lateral branches and shallow root systems that can extract moisture from the most meagre terrain or endure boggy waterlogged conditions and are among the planet's great survivors. They call to mind squat centurions. Dating from the Cretaceous period, the genus *Picea* has endured extreme heat, cold and a wide spread of atmospheres from oxygen poor to carbon dioxide rich. It appears to have spent the last ice age in refugia in the Rocky Mountains, spreading north as the planet warmed up.

In both Siberia and Alaska, neither the spruce nor the larch reach the Arctic Ocean. The mass of sea ice has so far kept the ground and the weather too cold. The ten-degree July isotherm marking the average summer temperature dips far south beyond the Bering Strait, describing a wide fan in the northern Pacific that was once the limit of the winter sea ice. The ice kept the coastal plains cool and the trees in check, especially on the Russian side, where the Oyashio Current flows down from the Arctic Ocean, cooling the coast and making Vladivostok the most southerly port to require ice breakers in winter. That is why, in the heart of the Chukotka peninsula, the easternmost district of Russia, which points like an arrow towards Alaska, the treeline does a ninety-degree turn south, tracing its way along the inside of the finger of Kamchatka around the edge of the Sea of Okhotsk. Climate breakdown is not closing the tundra gap on the Russian side. Instead, as we saw, the waterlogging of the soil is forcing the larch to retreat. But across the Bering Sea it is a different story.

The Alaskan treeline is made up of a mix of black and white

spruce, one loving well-drained dry soils (white) and the other seeking out the waterlogged bogs of the valley bottoms (black). This wide ecological niche gives the spruce a remarkably adaptive range. The spruce duo creeps up the Yukon delta and across the Seward peninsula always at a respectful distance from the shore, tracking the inlets, peninsulas and estuaries of western Alaska, only curving inland at the Kobuk and Noatak Rivers, where the forest meets the impenetrable wall of the Brooks Range. From the Kobuk watershed, the treeline follows the south side of the mountains all the way to the Canadian border.

Nobody knows this entire line as well as Ken Tape. For several years he spent his winters fighting through snow all along the FTE – the forest–tundra ecotone – up the Seward peninsula, around the edge of the Brooks Range on the Noatak River above Kotzebue to Point Hope, Barrow and Deadhorse on the North Slope, conducting a transect, a slice through the snowpack, to study its properties.

Ken was then a graduate student at the University of Alaska Fairbanks, researching the insulating properties of snow under the guidance of Dr Matthew Sturm. He was helping to prove Sturm's hypothesis that with warmer temperatures shrubs on the tundra trapped more snow, which insulated the ground, delaying the freeze and encouraging biological activity in the soil. This made nutrients available for plants, which encouraged further shrub growth and so on – a vegetative feedback loop that had the potential to change how climate change was modelled.[1] It was a controversial idea and one that took time to prove. Then, in 1999, they stumbled across the equivalent of a scientific treasure trove, big data from before the age of satellites that pushed their baseline way into the past: a cache of photos taken by the US Geological Survey in the 1940s.

This detailed survey of the entire North Slope of Alaska had been conducted as part of the mapping of the region for oil exploration. By re-photographing the region from the air, Matthew and Ken were able to compare the two sets of photographs and measure the differences. In fifty years the tundra had erupted with shrubs. The paper they published in *Nature* in 2001 was reported

around the world. It was the first significant scientific proof of a
process that Inuit elders had been speaking about for decades: the
Arctic was turning green.

Twenty years later, when I speak to him on the phone, Ken still
sounds like the youthful graduate student who made that landmark breakthrough many years prior. He no longer does fieldwork
in the winter, living in Fairbanks with his children these days, but
he still has the hunger and the excitement of discovery. I have
called at an important moment. He is about to publish another
groundbreaking study that, for the second time in his life, will
make headlines around the world.

He tells the story as though he is still surprised at his own dis-
coveries in a voice that – even on the phone – you can tell comes
from a mouth never far from cracking a smile. In the decades since
that first breakthrough discovery, Ken continued to study shrub
dynamics and observe changes on the tundra. But, he says, he spent
so much time with his 'head among the tussocks' that at first he
missed the bigger picture: 'the odd stuff that comes out of left field.
You have to keep your eyes open.'

Alaska is the most studied area of the Arctic; the US has the
resources and scientific heft that other nations lack. It is also sig-
nificantly different from Siberia, which is a more insulated
continental land mass; as we saw with Ary Mas, it takes many
years to warm up. Exposure to the warm currents of the Pacific
has meant the loss of sea ice in the Bering Strait, more abrupt
jumps in temperature and more apparent effects in Alaska. Plus,
the state boasts well-equipped field stations and scheduled flights.
In Alaska, the effects of climate change have been noticeable for
longer and more thoroughly documented than anywhere else on
earth. This northernmost state is a frontier in our understanding of
what is happening in geographic as well as scientific terms. The
largest scientific programme for studying the changes in the boreal
zone is run by the US National Aeronautics and Space Administration (NASA) and it is called the Arctic Boreal Vulnerability
Experiment (ABoVE). It aims to incorporate known research at all
levels of earth systems to try to model the changes under way.

'And yet we still know so little,' Ken says. Remote sensing only gets you so far. It must be supplemented by 'ground truth'. Yet sometimes it is easier to get data from space than observations on the ground. And what we see from one vantage point might look very different from another.

'It's hard to design a research project that is not run out of a field station.' Toolik Lake is in the Brooks Range on the North Slope and, along with another facility at Nome, is where most Arctic research takes place and where Ken used to spend a lot of time. At Toolik he studied other wildlife that had moved into the tundra from the forest due to the increase in shrubs and longer summers brought on by warming: snowshoe hares and changes in ptarmigan populations. Then he started wondering what might be coming next: moose, bears . . . beavers.

'Then I had this lightbulb moment that we might be able to detect these changes from satellites.' Most of the time with ecology, finding the animals and counting them is really hard, but beavers make such a footprint on the land, you can see it from space. 'I thought, we can track these guys. Oh boy, this could be big! But I'm not a beaver guy.' It didn't take much. Ken talked to a colleague who put him in touch with an expert, and soon they were surfing Google Earth, counting beaver ponds on the tundra.

He went back to the old aerial photos from World War II. No beaver ponds. The change was obvious and startling. Colleagues in Germany were studying changes to circum-Arctic thermokarst lakes – lakes that do not freeze because of dissolved methane – mapping the increasing surface areas of water, but they had not considered how the ponds were made. They had assumed melt. No one had realised that the landscape could be altered on such a scale by just one species.

'They were astonished, asking themselves, how did we miss this?' says Ken.

The research station at Toolik Lake is reached by the Dalton Highway, sometimes called the haul road, the only paved road in northern Alaska. It keeps the Trans-Alaska Pipeline company from Prudhoe Bay through the Brooks Range to Fairbanks in the middle of the state and all the way to the port of Anchorage in the south. The transition zone from tundra to forest along the road in the mountains is very narrow, a matter of feet rather than miles. Trees stop on one side of the valley and then do not reappear. The gradient is steep. There is not the broad, diverse interaction between forest and tundra like in Siberia or Canada or the tundra plains of western Alaska that face Russia across the Bering Strait. There is a tree called the LAST SPRUCE that someone has marked with a sign by the roadside on the North Slope, but it is an outlier, probably from a seed that hitched a ride on an oil truck.

'The beaver effect was not really apparent from the haul road; that's why we hadn't noticed.'

But once Ken started looking, he saw beavers everywhere. He counted ninety beaver ponds on the Baldwin peninsula, up from two dams twenty years before. In total he found 12,000–13,000 beaver dams on the tundra that were not there in 1950. Ken did some modelling that led to a hypothesis and scientific paper that as we speak is under review, and which, when it is published a few months later, is reported in newspapers all over the world. Beavers, it turns out, have more impact on surface water in Alaska than climate. They could control up to 66 per cent of surface water on the tundra, paving the way for trees.[2]

Beavers don't need a full forest to flourish; the enhanced shrub growth on the tundra of the last thirty to forty years, dwarf willows and alder are enough for them to build dams and make ponds. Water conducts heat better than land so when you create more water and make it deeper, you are making the environment warmer, and you are carrying the warmth closer to the soil, to the permafrost. Beaver ponds create a foothold for more trees and for other species that rely on them: amphibians, insects, fish and birds. They are geo-engineers of the first order.

So the forest–tundra ecotone in Alaska has a new keystone species, *Castor canadensis*. The North American beaver is much more numerous than its Eurasian counterpart, *Castor fiber*. The Eurasian beaver was almost extinct in Europe and Asia when the Soviet government introduced protection in 1922, and even now reintroduced populations have only resulted in small numbers in southern Siberia and east of the Urals. The North American species by contrast has rebounded and is now over five million strong and growing.

I press Ken for tips on where and how I can see the 'beaver effect' for myself. He suggests looking at the tundra beyond the treeline to the west of the Brooks Range. Beavers have not been seen on the North Slope yet, although they are in the mountains and in some of the passes that lead to the other side. It won't be long. The tundra of western Alaska is a preview of what the North Slope is going to look like when they get there.

Ken himself is planning research later in the summer of 2020. It is the first full field season on the project, and he has a team ready to install cameras, collect data on fish, mercury, aquatic food webs and the impact on indigenous subsistence hunting and trapping. I am excited about the possibility of joining this groundbreaking trip, although Ken is non-committal. He suggests I look around the Kobuk River, and mentions a writer called Seth Kantner based in Kotzebue who grew up on the river. We chat for a while about access, how to get there, the difficulty (and expense) of bush flights. Northern Alaska is, apart from the Dalton Highway, a country without roads. We promise to meet up when I get to Fairbanks. But then, a few months later, the coronavirus pandemic strikes.

Kotzebue, Alaska
66° 53' 53" N

I phone Seth Kantner and ask about the possibility of visiting Kotzebue.

'Ah, I don't know,' says Seth. 'What kind of a guy are you? I mean, are you a white guy?' He gives a small laugh. I laugh too.

'Well, then, it might be tough.'

The last time a global pandemic swept through Alaska, in 1919, it nearly wiped out the native Inuit, Iñupiat and Athabaskan populations. There are old stories, Seth writes in his memoir *Shopping for Porcupine*, of 'famine and influenza – dog-team travellers finding half-starved children in igloos full of frozen dead people'.[3] The history of the frontier is still shaping the present: fearful native councils, the local municipal authorities, have instituted their own stricter lockdowns and restrictions on visitors including scientists and journalists and so I must grapple with the remote sensor's dilemma: how to understand what is happening on earth with only the view from space.

Seth himself is white but, he says, and everyone knows, that is only in a manner of speaking. The rules don't really apply to him. Seth grew up, as he puts it, 'more native than many native kids' in a sod igloo built by his parents on a magnificent bend of the Kobuk River miles from any other settlement. The view took in the fringe of the forest to the south, a treeline of spread-out, spindly, stunted spruce striping the snow. This was where in 1965, fresh out of college and keen to avoid regular jobs, his parents Howard and Erna Kantner swapped the banalities of a conventional modern life for the wilderness and a subsistence existence, living off the land.

The Kantners were part of a movement of people from the 'lower 48' seeking a simpler life, a movement the *New Yorker* journalist John McPhee wrote about in his 1976 book, *Coming into the Country*. Seth and his brother took correspondence classes from books that arrived at their igloo via dog sled from the post office in Ambler several days' journey away. Their true education, however, was in the struggle for survival in tune with the seasonal cycles of the land. Fishing in summer and trapping in winter. They learned to make fish nets and traps from spruce roots and willow bark, sleds and snowshoes from birch; they knew where to find berries in autumn and learned to track, shoot and skin caribou, moose, bear, fox, muskrat and wolf, and how to make clothes from their fur, hides and sinews. Seth discovered the best way to find a camp spot was to follow fox tracks to willows. Now you can't move for willows, and not just in the creeks.

In 1975, when Seth was growing up on the Kobuk, John McPhee canoed down the Salmon River, a tributary that joins the Kobuk a little way downstream from Seth's family place. He accompanied officers from the National Park Service, the Bureau of Land Management and the Sierra Club who were surveying the land in the wake of the Alaska Native Claims Settlement Act (ANCSA) to assess its conservation potential for inclusion in a series of proposed national parks. They were, McPhee wrote, 'legionaries from another world ... Romans inspecting transalpine Gaul'. The question with which they and McPhee were concerned was, 'What is to be the fate of all this land?'

Fifty years later, we are beginning to learn the answer. ANCSA awarded the original inhabitants of Alaska a billion dollars and 44 million acres of land in exchange for giving up any further land claims. It paved the way for the Trans-Alaska Pipeline and demarcated a further 80 million acres as 'national interest' lands, to be studied for their conservation potential, eventually resulting in 32 million acres of new national parks, an area the size of New York State and more than all the other US national parks combined. When President Richard Nixon signed it in 1971, ANCSA was arguably the largest, most progressive bargain with indigenous people in history; an attempt to square the competing demands of the twentieth century: oil, conservation and the end of the colonial era. Alaska therefore represents in microcosm the tragedy of modern industrial society. Despite the best efforts of the richest country on earth to do the right thing on conservation and indigenous rights, the enduring commitment to the third demand – hydrocarbons – is undermining the viability of the rest.

McPhee does not mention climate change but, as we now know, the processes were already well under way. Seth is fifty years old and global warming has been the backdrop to his entire life.

'The old Eskimos would say you couldn't camp much north of our house,' he tells me on the phone. Camping required two willows to bend together for a shelter, and wood to burn. There were never many willows north of Paungaqtaugruk, their bend in the river, and no spruce to speak of. In *Shopping for Porcupine* Seth visits his parents' old house and finds the path sunk by two feet and the entire front of the hill dropped from permafrost melting. The view of the tundra is confusing so he hunts through drawers for an old photograph to compare how it used to look. The photo is in the book. It shows Seth's parents and their friends in 1965, ruddy-faced, idealistic pioneers, the endless flat tundra stretching to grey moody mountains in the distance. Beside it is another photo – of the same landscape forty years later, a surging mass of green fingers reaching to the sky, a field of 'happy spruce', Seth calls them. Climate change is as old as he is; older, in fact.

'White spruce is coming up like grass ... you look away and they move!'

The eruption of vegetation makes navigating familiar terrain difficult even for seasoned hunters and trappers who have grown up on the land, like Seth.

'You drive along a river and you can hardly tell where you are any more,' he says.

'It's shocking for all of us. We always considered ourselves at treeline here.' A small chuckle escapes. 'The so-called treeline.'

The trees bring other species: birds, moose, bears and extraordinary numbers of fish, fleeing warmer waters elsewhere. The salmon run is haywire. Seth has been fishing commercially since he was nine. Things are a long way out of whack. Three years ago fishermen caught 100,000 salmon on the Kobuk, two years ago 200,000 and last year 500,000. Salmon have become refugees. But sometimes this usually cold river is still not cool enough for salmon seeking a safe haven to spawn. In 2014 and 2019 the river had more dead fish than living. The water temperature stayed above seventy degrees Fahrenheit for a month. An Alaska Department of Fish and Game plane that flew for 200 miles along the river posted pictures of the bloated bodies of salmon collected like driftwood in the shallows and bears feasting. The banks were piled high with carcasses, roe and blood strewn over gravel like streaks of orange and red paint. The brown bear population is expanding fast, and the food chain shifts accordingly as the bears push out other predators.

The whole ecosystem has been undergoing a slow-motion

transformation for more than fifty years, and yet those people who used to rely on that ecosystem are not rioting in the streets. Instead, they adapt.

Seth no longer lives upriver but in the port of Kotzebue that sits on a narrow spit, a few metres above sea level in the middle of a huge bay called Kotzebue Sound. The town is arranged around the eroding shoreline of the land spit. Small shacks, jetties and boats pulled above the waterline crowd the shingle beach in front of the $34-million sea defences that hold up Shore Avenue. It looks north-west to the sea, out to the rich fishing grounds of the Bering Strait, from where, traditionally, the indigenous Iñupiat have always sourced their food. Beluga whales were a staple of their diet alongside seal for thousands of years, but the abundant multi-tudes of beluga in Kotzebue Sound are a folk memory now. No one knows why these highly intelligent animals have stopped returning.

'Twenty years ago it was politically dangerous to talk about global warming,' he tells me, 'but these days everyone accepts it; it's just a fact of life.' The changes are so obvious now that denying them is pointless. The ice in Kotzebue Sound is no longer reliable in autumn and spring. More than a decade ago, after several fami-lies lost loved ones when snowmobiles went through sea ice at times of the year when it had previously been strong, old winter routes were abandoned. It is a glimpse perhaps of the coming era of climate politics in other places. Our evolutionary success is also our strategic weakness.

Humans have been living with the degradation of their environ-ment for so long in Kotzebue it is no longer remarkable. Unable to visit, I read instead the accumulated media reports of two decades of climate coverage of Kotzebue and was struck by the overwhelm-ing feeling of resignation, the Iñupiat mourning the passing of a way of life, accepting their own powerlessness with a grace that is heartbreaking. When the whales go, they hunt seal. When the seal go, they shrug their shoulders and turn instead for help to the gov-ernment whose easy oil money is both the source of the problem and the immediate solution.

Oil exploration leases have included dividends to indigenous corporations, and the structure of the old life of Alaska, like the permafrost, has been fatally undermined. Subsistence, the ideal for which Seth's parents moved to the Kobuk valley, is no longer a viable proposition. Not because nature has stopped giving – not yet – but because the people have stopped harvesting. Seth's parents' ban on shop-bought processed food has not survived a generation. Instead the oil and the imported consumer goods have meant the cost of living has shot up, and people have come to depend on welfare cheques, free housing, shotgun ammunition and now stimulus dollars. This, and the 'sickness' as one of Kotzebue's elders puts it, of social media, have overtaken the old ways. Hydrocarbon culture has suffocated the keen beauty of survival driven by hunger. The hunger is not missed, and now there is no way back. The knowledge of how to survive is still there, still alive, just, carried by people whose lives have seen more profound change than almost any other lifetime. But it is going.

'Almost everything comes off the airplanes now,' says Seth, a note of disappointment in his voice. In the course of his own lifetime, eyes and ears that once scanned the tundra for the signs of caribou, geese, wolf or a dog team now wait for the buzz of propellers and the ping of a phone for their food and news.

Agashashak River,
Kobuk Valley National Park, Alaska
67° 34' 92" N

From space, satellites confirmed Ken Tape's picture of Alaska getting greener. Over the last two decades this led to models and predictions of the treeline leaping north. However, things are not turning out quite as expected.

Seth connects me to a pair of scientists he has got to know over the years who pass through Kotzebue every summer on their way inland to a field site where they are monitoring the growth of white spruce. When I call, Roman Dial and his colleague Paddy Sullivan are frustrated. For fifteen summers they have been returning to the same spot on the treeline, a remote valley on the Agashashak River, a tributary of the Noatak, in Kobuk Valley National Park. In 1979 Roman went rock climbing in the Aregache Peaks, camping above the treeline. There he ran into a botanist from Colorado State University called David Cooper, who was looking for white spruce hundreds of metres above where they should have been. Roman joined in and was hooked. So began a lifelong obsession with the treeline.

Paddy and Roman's ABoVE project is one of the longest-running ecological experiments in the Arctic and has yielded valuable data about changing vegetation dynamics, particularly the fate of spruce in Alaska. But this summer they will not go. They will miss a whole year's worth of data, compromising the science.

The thing that exercises Paddy the most is fertiliser. Every year at Agashashak he fertilises a patch of white spruce with nutrients next to a control patch without fertiliser. The aim is to understand what limits the growth of trees at extreme latitudes and altitudes. One of the remaining mysteries of the changes happening in the treeline zone is why in some places the forest is racing north (a hundred metres per year in Scandinavia) and in others it is taking its time (less than ten metres a year in central Canada). Computer models based on temperature alone predict rapid forest advance in most places, but the reality is more subtle as different species respond to warming in different ways. Alaska contains examples of both advance and stasis in one species, and this is what Paddy and Roman are studying.

East of the haul road white spruce appear to be going nowhere, but in the lowland more maritime west they are racing into the tundra at speed as if running for their lives from the fire and drought further south. If Paddy's team can understand the limiting factors of this keystone species, they can better predict what the future landscape might look like, and how much carbon it may sequester or how much radiation it might absorb. Missing a year of fertiliser will screw things up.

Piecing together a remote picture of the Agashashak valley I 'fly' upriver from Kotzebue like an old-world shaman travelling the landscape in a dream. Behind the spit of land where Kotzebue sits, the Kobuk and Noatak dissolve into Kotzebue Sound in a flat maze of delta. Here the treeline meets the saltwater in an eerie landcape of yellow sand that looks more like the Sahara than the Arctic. The southern bank of the Kobuk delta is made up of 200,000 acres of sand, the remains of glacial morraine blown into the sheltered valley. Powdery golden aeolian sand is common in the far north. This is where two geological forces at work since the last ice age have met: retreating glaciers and advancing trees. The lee in the land that makes it a favourable spot for trees also means it is where the wind deposits sand borne across the plains. The dune-scape is spooky: huge undulating yellow and purple shadows that evoke associations of heat not cold, of desert succulents and camels, not the occasional flower of a pink fireweed poking out of the sand or a stunted spruce standing, surprised, against an immense backdrop of dunes. It's almost as if they know they're not supposed to be there.

Up the river delta the threads of the sea-bound creeks emerge from the sheets of mud and sand to coalesce into the main stream and begin their meander that is the driver of the forest's rhythmic succession. From the air, stripes of green follow the bends of the river like ripples radiating outwards. The forest seems as if it has been painted in S-bends again and again on the surface of the plain, and at a lower angle the varying heights and densities of the trees become apparent, each stripe, each line different, creating an effect like fabric, of threads laid down repeatedly and pulled tight against each other.

As the currents of the river carve away at the outer banks of the bends they mine the ground from under the mixed spruce forest – white and black, old growth that is by now 150 to 200 years old in what is called a climax state. Along the inner banks the water deposits a train of gravel which is slowly being colonised by willows, grass and moss. Beyond is a curve of higher willows, alder and birch running parallel with the river, and behind this band of

youthful forest is a more open wood mostly consisting of the narrow spires of black spruce and an occasional, taller, white spruce.
The black prefer the moist boggy soil of the river plain while the
white like the bedrock of the valley sides.

As the meanders progress up the valley the forest succession follows in parallel. It is a picture that has been painted very slowly
over thousands of years by the deliberate brush of the river.

Along the valley bottom the well-drained soils favour the trees,
but at the edge, behind the fringe of spruce up the valley side, the
permafrost layer impedes drainage and lowland tundra flourishes:
lichens, mosses, berries, and the standing water of lakes and ponds
in a mosaic of tussocks. Higher again, where the bedrock pokes
through a thin layer of soil, the spruce resume. It is what is called
an inverted treeline. The dense forest above, alpine tundra below
and a band of forest tracing the route of the river. More recently,
runs of cottonwood (poplar) have started to appear, starting out as
shrubs and shooting like streaks of flame up the valleys in autumn.

'Yeah, the cottonwoods are going crazy!' says Roman. But
simply going to the same site year after year is not enough. And
this is the difficulty for anyone trying to attune to tree time, or
earth time. 'It's like watching a kid grow up – you don't notice,'
says Roman, who hopes to be a grandfather soon.

They looked at earlier photos. When they compared their plots
with aerial photos from the 1950s, they found the forest had
expanded only a little, not as much as the models predicted. The
real shock was that a previously marginal ecosystem – tall shrubs
with occasional trees like the cottonwoods – had increased by 400
per cent. Warming is shifting the balance of the ecosystems at the
treeline, the transition zone appearing to expand in favour of
shrubs not forest.[4]

The greening is not leading to the advance of the treeline, but
the evolution of a whole new landscape for Alaska. What's more,
the tundra as a whole is boiling up with vegetation. It's not a question of a speed-up in the advance of a line of trees or species. When
you light a piece of paper with a match, a glowing red ember creeps
across the page, but when you put a sheet of paper in a furnace, the

whole lot ignites at once. 'Tundra' will soon be a historical term in Alaska. Boreal species of birds and shrubs are already present in the Arctic National Wildlife Refuge on the North Slope, says Roman.

Every research question ends with another question. Paddy and Roman wanted to know what was going on with the shrubs: what did the willows and the alder have that the spruce did not? And might the answer help explain the difference between the spruce in the east and the west of the Brooks Range?

'We only have seventy-eight data points,' says Paddy, explaining how little they know about why some trees flourish and others die as a result of warming. 'We know nothing.'

With the help of an expert in the forest underground, Rebecca Hewitt, they began looking at the fungal associations of spruce.

Spruce has a strong reliance on mycorrhizal networks and on lichen to supply it with nitrogen and minerals, as most plants do. Over 90 per cent of plant species rely on fungi to survive. One end of a fungal fibre embeds itself in or around a tree root. The other end of the fungus, a thread called a hypha, can be fifty times finer than the thinnest roots and hundreds of times longer. This effectively extends the reach of the tree's root. Globally, these threads of mycorrhizal fungi make up between a third and a half of the living mass of soils.[5] Soil is in fact a huge, fragile tangle of tiny connected threads.

In the eastern Brooks Range, in the drier, colder, continental climate, Paddy and Roman found that fungal activity was high, even in very cold soils with continuous permafrost. The trees were neither growing nor advancing east of the Dalton Highway but they were investing heavily in their fungal partners in order to survive. But in the western Brooks, where the maritime influence of the Bering Strait makes things warmer and wetter, there were fewer fungal associations yet growth and advance was more pronounced. The assumption they made was that the trees didn't have to invest so much in fungal relationships to access the nutrients they need.

Rebecca Hewitt has striking red hair and blue eyes which contain the steady glow of the seeker. A patient communicator, on our video call she slowly and clearly explains to me how nitrogen is needed by plants to build their photosynthetic machinery, how the tundra and the boreal is nitrogen limited, and how, when this critical element arrives, the tundra is overwhelmed by vegetation. She has observed plants in the tundra that seem to pull nitrogen from deep underground and then, through complex symbiotic relationships mediated by fungi, share the nitrogen and various minerals with other species. The nitrogen-fixing plants themselves do not exhibit signs of enhanced growth, while others such as willows do very well. Where is the nitrogen coming from? Could the plants be mining it from the degrading permafrost? Who is sharing what with whom in this tangled web of exchange beneath the surface of the soil? And who decides the balance of resources, of power?

We know that fungal networks can transport carbon, water and minerals between trees and other plants, even across species. The idea of the 'wood wide web', stemming from ecologist Suzanne Simard's pioneering research into birch sharing carbon with Douglas fir in the Pacific Northwest region of North America, has gained notoriety.[6] Studies since have shown hundreds of kilos of carbon in a single hectare being moved between trees through fungal networks. However, the concept of fungal networks as passive conductors of material and information is, according to mycologist Merlin Sheldrake, a very 'plant-centric view'. Fungi, he reminds us, have their own interests. He suggests a better way to think about the role of mycelium in the soil may be as brokers. But listening to Roman and Paddy talk about the explosive response of the shrub layer to warming at the expense of trees, a different analogy springs to my mind. What if the fungus is more like a farmer (of carbon and sugar) diversifying their crops in response to an uncertain climate?

It could be that processes of decomposition, soil formation and fungal migration are just as important as temperature rise in driving the large-scale vegetational shifts currently changing the surface of the earth. It would make neat spiritual as well as biological sense: that the pattern of life is shaped by the pattern of the life that went before.

We are only just beginning to peep below the surface of the forest floor. We still have very little idea about what is going on. All we do know is that the forest, like all life, is a symbiotic system, a dynamic process, not a collection of things or distinct beings. And the more closely we examine it, the more mysterious it becomes. Somewhere down there among the web of hyphae and root tips and the permafrost could be the frontier of the forest. It seems that the key to understanding where and what the treeline is and how it will respond to warming and melting is not to be found among the greening spires of the eager spruce above ground, but in the humid black organic layer below.

The greening of the tundra, however, is only half the satellite picture. Further south the forest is turning brown.[7] Rebecca's friend and colleague Brendan Rogers is a climate modeller based at the Woods Hole facility in Massachusetts and involved with many of the ABoVE projects, particularly estimating carbon stocks in soils and forests. He explains on the phone that at first small-scale satellite imagery did not pick up the browning of individual trees within stands of healthy forest, but by 2010 the picture was beginning to shift and by 2020 a process of widespread forest decline seemed well under way. Satellite photos from space of the North American boreal today reveal green valleys, black burn scars and whole swathes of brown – heat-stressed or insect-damaged spruce. Instead of an even carpet of pulsing livid green, the earth appears to have a skin disease.

In a way, the greening and browning are linked, Brendan says: 'Warming was expected to increase the productivity of the forest,' as growing seasons lengthened and plants photosynthesised more. But the so-called carbon dioxide fertilisation effect was short-lived.[8] NASA scientists discovered a moisture and nutrient limitation to photosynthesis as well as a temperature one. Warmer air holds more water vapour, increasing the vapour pressure deficit, the mechanism which encourages plants to transpire and release water vapour into the atmosphere. Essentially, warmer air sucks more moisture out of leaves. In order to avoid losing moisture in warmer temperatures, trees are closing up their stomata and stopping photosynthesis. Even if there is plenty of water in the ground, if the rate at which they are losing water exceeds the rate at which they can take it up then a tree will do the sensible thing and shut down growth, limiting its ability to build leaves and to sequester carbon.[9]

Spruce needles are rolled-up leaves covered with a waxy cuticle that prevents excess moisture loss, and each stomata is also modified to conserve water. This resilient conifer of the northern woods has evolved to survive with very little moisture, but it cannot do without water completely.[10] It is amazing to consider how very little we know about how trees die. When the fine needles on the tips of the spruce's branches turn brown and fall, as they do increasingly every summer, a coroner would struggle to determine the cause of death. Is it water deprivation, vascular damage or carbon starvation?[11]

Spruce are tough, their xylem vulnerability to water stress low, which is why they are prized for pulp. The little tracheids that plumb and feed the growing tree are strong and thick-walled. These are what make the fibres of paper. But extreme heat puts both white and black spruce under pressure. Winter or summer drought is the same: lack of water in the tree increases the tension of the column. Spruce have evolved a unique mechanism for managing this tension, which can rise as high as 900 pounds per square inch – a car tyre maxes out at 40. They have small antechambers between the cells that connect to each other and maintain the integrity of the whole while allowing flexibility of the moisture composition within each cell. There seems to be no lower limit for freezing, but in warm weather even the spruce has its breaking point. Without water in its roots, the pressure of the water being slowly consumed in the tree causes tension, pulling the water up the xylem from the roots. Eventually, the xylem draws air bubbles into the cell walls, causing xylem cavitation. When the bubbles join up it is akin to an embolism in a person's veins, like a diver getting 'the bends'.

Like most northern species, the spruce has evolved to lie

dormant for most of the year and take advantage of the short
growing season to build reserves for the harsh winter. But if it must
spend the summer dormant as well, surviving heat or drought
stress, then it has little opportunity to grow at all. Summer growth
is becoming a luxury.

At the same time, this acceleration in the water cycle means
there is more transpiration from all vegetation, which increases the
convective energy in the system. This leads to more storms, more
thunder and lightning, which means more ignitions and more fires,
the doubling of burned area and exponential increases in emissions nearly every year. In 2019, Brendan says, Alaskan forest fires
emitted seventy terragrams of carbon dioxide, the same as human
activity in the state of Florida.

Spruce don't need to be dead to burn well. Black spruce is called
'gasoline on a stick' by firefighters because of its flammable gum.
The leaf litter of spruce contains camphor – used in pyrotechnics.
This is because black spruce only regenerates after fire: its sticky
black serotinous cones lie dormant until the gum is melted, releasing the seeds. But with larger and larger areas burning every year,
as in Russia, traditional patterns of succession are becoming disrupted, and black spruce are not regenerating.[12]

'The boreal forest is breaking apart,' one Alaskan researcher
said back in 2016. 'You lose spruce and you lose everything that
lives in spruce, and that is basically the boreal forest.'[13] This is not
just a problem for the inhabitants of the forest, but for all of us.
The feedbacks and interrelationships between forest systems,
water cycles, atmospheric circulation, carbon storage and permafrost melt are complex and far-reaching, too complex for any one
computer model alone.

'All we can say for sure is that there is going to be a lot more
climate disruption,' says Brendan. To glimpse what might change
when we lose a forest, you first need to understand the role they
play in maintaining the status quo.

An intact spruce forest, like any forest, is principally concerned
with creating and maintaining its own habitat. We know that trees

make rain. Spruce are particularly good at this, with potent volatile organic compounds that bond to molecules of water vapour, condensing them into water, making them heavier, so they fall as rain. What's more, the trees also put the water vapour there in the first place. Each tree is a tiny independent rain-making factory. Trees take up and transpire far more water than they use for photosynthesis; up to 90 per cent of the water they take in is not used. Why do they do it? As the scientific journalist Fred Pearce puts it, 'Trees release moisture to make a world fit for more trees,' and us.[14]

Fifty per cent of the rain falling on land originates from evapotranspiration in trees. And since trees continue to take up and emit water, recycling the rain that they make, contiguous stretches of forest seem to be important highways for rain and wind, as rain that falls on forest is transpired and then falls as rain again across continents, like a kind of pump that has been called 'flying rivers'.[15] Other research has called this phenomenon 'teleconnections' between forests on different continents, such as the link between the Amazon rainforest and the west African monsoon. The spruce forests of Alaska and northern Canada seem to have a direct relationship with the rainfall in America's bread basket, the great plains of the American Midwest.[16] Research on teleconnections is just beginning, but the same appears to be true for the Russian taiga and the wheat fields of Ukraine.[17]

Forests also help make the wind, although there is still debate about how. The polar front is the sharp junction between the mass of cold air over the North Pole and the warmer air from temperate latitudes. It moves with the seasons. In winter it often dips down into lower latitudes, bringing with it ice and snow, but in summer it has usually been stable, occupying a position more or less aligned with the treeline. Until the 1990s it was assumed that the position of the trees was influenced by the wind, but research by Roger Pielke and Piers Vidale, a pair of UK-based scientists, suggested the reverse might be true: that the trees decided the position of the front.

Spruce trees are dark green because of the highly concentrated mesophyll tissue in their needles, which absorbs radiation. As with the high pines of Scotland, this colour can be so dense and the waxy cuticle on the needles so thick that the needles appear blue or almost black. When sunlight hits the spruce's needles, the microstructure of the tree – its stems and needles – bounces the radiation between them, absorbing the short-wave radiation and making the waves longer, ensuring that as many photons as possible are collected. Because the trees absorb light, less infrared radiation is released back into the atmosphere to be trapped by the carbon dioxide blanket and converted into heat to fuel global warming. The great spruce forests of North America not only produce vast amounts of oxygen and absorb carbon, they cool the planet while doing so. One tree in summer contributes 70 kilowatts of cooling for every 100 litres of water transpired, equivalent to two conventional air-conditioning units.[18]

In the winter this ability to absorb radiation is the reason snow will melt around the trunks of spruce trees. They warm the soil around them and contribute to a conducive habitat for insects, rodents and fungi in the sub-nivean world. The temperatures between spruce branches and beneath the trees can be considerably higher than the air beyond, as the tree re-radiates energy absorbed from light downwards to the snowpack. This is why human inhabitants of the northern forests often camp beneath spruce trees and hold large specimens sacred as places of shelter.

But in summer the dark colour of the trees leads to a huge difference between the reflective capacity, or albedo, of the tundra and the albedo of the forest. Because the boreal forest absorbs so much radiation, it heats up like a black road surface in the sunshine. The neighbouring tundra by comparison reflects most radiation straight back out into space. The difference between the two is of such a scale that it creates a steep temperature gradient between the tundra and the forest. This gradient, Pielke and Vidale proposed, drives the winds and determines the position of the polar front.[19]

Subsequent research however has suggested that temperature gradients are not the whole story. More than a decade ago a Russian physicist called Anastasia Makarieva proposed a novel theory

that by the time of the ABoVE conference in 2020 had started to
gain more attention.

Makarieva was interested in the vacuum created when the water
vapour released by transpiring trees cooled and condensed into
water droplets. As a physicist, she started with the principle
that water, a liquid, takes up much less space than water vapour, a
gas. The transformation from gas to liquid creates a partial vacuum,
a pressure drop, into which more moisture-laden air from below is
sucked up. This rising wet air is then replaced lower down by air
moving in horizontally over the forest canopy. Makarieva and her
colleague Viktor Gorshkov called their theory the biotic pump. The
process of making rain also pumps air over the forest to make the
rain move along. This wind will keep moving as long as there is
water vapour condensing, so as long as the trees are transpiring.[20]

The notion of a biotic pump might help explain why the position of the polar front and the treeline correlate in summer (when
the trees are transpiring) and not in winter (when they are dormant). The spirit of the north wind – *boreas* in Greek – could it
seems reside in the north woods, the boreal. Makarieva told Fred
Pearce that forests 'are not just the atmosphere's lungs, they are its
beating heart too. The biotic pump is the major driver of atmospheric circulation on earth.'[21]

If Makarieva is correct, then forests should be considered
natural assets of the utmost importance, vital not just for the main-
tenance of human habitats in the places where they are, but
biological engines with geopolitical relevance across borders and
continents. As more attention is paid to the science of how our climate is changing, and as more questions are asked once the
'ecosystem services' that we have taken for granted start to falter,
could it be that nation-states will start taking an interest in deforestation in foreign countries, and instead of dispatching armies to
maintain the flow of oil launch invasions to protect the forests that
make and send the rain?

The biotic pump theory also seems to explain the changes taking place in Alaska's forests right now. As forests transpire less,
they make less rain and less wind, leading in turn to hotter and

drier conditions more favourable to drought and fire. Summer winds in Alaska already appear to be weakening, creating longer, drier high-pressure systems that linger, accelerating the drought.[22] The boreal forest is foundational to the climate system of the last few million years, and as the trees march towards the sea, dry out or are obliterated in a shrubby mass of greening tundra, the stable winds that purred away at the top of the world, quietly regulating the northern hemisphere, will go – are already going – have gone – haywire.

Huslia, Koyukuk, Alaska
65° 42' 7" N

Solastalgia is a neologism that is becoming commonplace. It describes a feeling of homesickness while still being at home. It is a sense of loss, but also a feeling of confusion: the planet that we think we live on no longer exists. In a warming world, words become unmoored from their meanings. The anthropologist's gap between the signifier and the signified opens up like a dangerous crevasse. 'Tundra' no longer accurately describes what the tundra has become. 'Spring', 'winter' and 'autumn' will soon be contested concepts in many parts of the world. While such feelings will shortly be experienced by most humans, those who were close to the land have been living with them for nearly a century already.

In the future the history of oil in Alaska may come to be seen as a terrible irony. When John McPhee and his colleagues went down the Salmon River in 1975 it was oil, ultimately, that set them on their quest. The discovery of oil led to the Alaska Native Claims Settlement Act. And it was outrage at the oil pipeline that spurred the inclusion of provisions about 'national interest' lands and the National Parks Service to send out rangers and scouts to survey and assess the land from the Arctic Ocean to the Pacific and all the great rivers and mountains in between.

A major element of the National Park Service's programme was the Cooperative Park Studies Unit established at the University of Alaska Fairbanks under anthropologist Zorro Bradley. Bradley recruited a team of researchers to document in great detail the 'life-ways' of native populations living in areas being considered for national interest status. The NPS was committed to managing any future national parks in line with the cultural and socio-economic interests of the people who called these habitats home. It was a path-breaking programme that yielded a rich inventory of what, forty years later, is steadily being lost due to the grand bargain with private property and carbon-fuelled consumption represented by ANCSA. The effort was valuable for another reason: the record of the indigenous people's wisdom and world view instigated by the oil strike also holds much promise for a way out of the maze of confusion and denial that the oil has built.

At the headwaters of the Kobuk the watershed stops at a continental divide, a low ridge of blue hills running north into the wall of the Brooks Range. Rain falling on one side of the ridge travels down the Kobuk to Kotzebue and the Chukchi Sea. Rain falling on the other becomes the Alatna then the Koyukuk and eventually the Yukon, reaching the Pacific Ocean several thousand miles to the south. When Seth, the writer who grew up in an igloo on the banks of the Kobuk, was young, the continental divide was the beginning of the forest proper. The people of the Koyukuk River, known as the Koyukon, are, unlike the Iñupiat with their eyes on the sea, entirely a forest people. The same year that McPhee went down the river, in 1975, a thirty-five-year-old anthropologist and his team of dogs entered Koyukon territory, crossing fresh spring snow. It led to an encounter with profound consequences both for the anthropologist and the Koyukon people.

Richard K. Nelson had had his first taste of Alaska working for the US Air Force in Wainwright at the age of twenty-two. After that he had studied anthropology and become a lecturer on life in the Arctic in Hawaii. In 1974 he seized the chance to work for Zorro Bradley's new Cooperative Park Studies Unit, and returned to

Alaska. Bradley sent him to Ambler and Shungnak, Seth's old stamping ground along the Kobuk valley. After a time, he headed east, inland, to the Koyukuk to spend a year at Huslia, a year that would in time be transformed into one of the most significant works of post-war ethnography, a classic that inspired a television series and is still in print: *Make Prayers to the Raven: A Koyukon View of the Northern Forest* (1983).

Nelson's key insight and achievement was to take the Koyukun 'lifeway', their world view, on its own terms and describe it in his. *Make Prayers to the Raven* is, he wrote, 'a native natural history that stands outside of the realm of Western science'. The book contains exhaustive appendices of Koyukon names for species and concepts, an important contribution in its own right, but more than that, Nelson shows how the Koyukon view of nature is tangible and above all real, even if it is beyond our emotional grasp. He wrote,

'The Koyukon people live in a world that watches, in a forest of eyes. A person moving through nature – however wild, remote, even desolate the place may be – is never truly alone. The surroundings are aware, sensate, personified. They feel. They can be offended. And they must, at every moment, be treated with proper respect.'[23]

For the Koyukon, this world view shapes how they live and survive in the land. It is a glimpse of what it might mean to truly live as part of an ecosystem: 'Ideology is a fundamental part of subsistence . . . most interactions with natural entities are governed in some way by a moral code that maintains a proper spiritual balance between the human and nonhuman.' For the Koyukon, like many other indigenous communities, humanity, nature and the supernatural are joined together in a single moral order in 'the world that Raven made'.

It is an inspiring vantage point with so much to offer our current hydrocarbon bind. The book made me see landscapes anew. After finishing it and armed with fistfuls of questions I went in search of Richard K. Nelson, who had gone on to become a writer of distinction after his time in Huslia, serving as the laureate of

Alaska as well as a familiar voice on US National Public Radio with a long-running series on soundscapes and nature appropriately called *Encounters*. I discovered with sorrow that he had died weeks before, while I was reading his book. After his life-support machine was turned off, he had requested to be left alone with a recording of a raven's call.

So I searched for his main Koyukon teacher, Catherine Attla, and found that she had passed away a few years before. She had also been a household name on Alaskan Public Radio, hosting a show collecting and sharing indigenous stories and knowledge called *Raven Time*. The archive of these recordings at the University of Alaska Fairbanks is a rich resource for a new category of history, the oral history of global warming.[24]

K'ititaalkkaanee, as Catherine is known in the Koyukon language, appears on the screen, a dark-haired woman dressed in a warm parka with a ruff of wolf fur sitting outside; above her spare birch branches frame a pale blue sky. The sun is on her smooth face, glinting on the steel arms of her spectacles. She is smiling as though laughing at her own story.

'Shhh!' she says, giggling. 'That's what they used to say: "You must be quiet in the presence of the ice. You must show it respect."'

It is springtime. The ice is going out in the river: great slabs of shifting grey plates jostling and cracking and tumbling at a rate of knots downstream, grinding past the village of Huslia. The sun shines on spindly birch trees, spires of bronze spruce and square houses with low pitched roofs made of spruce logs. A group gathers by the riverbank in caps, sunglasses and fur-trimmed parkas to offer Christian prayers and sing traditional songs, giving thanks to the ice: 'May the same number of us see you move again next year, oh River Ice.'

Catherine tells how when she and her sister threw sticks onto the frozen river the elders would tell her off. There is a spirit in the ice. And it is powerful. They were forever being told to keep quiet and not to speak of 'big things you don't understand; don't talk big, your mouth is small!' They were forbidden to talk about the sun, the moon, the sky, the animals. The world is animate, it can hear. And if an animal doesn't like being spoken about, it will bring you bad luck.

'I can't wait to be big,' her sister used to say, 'so I can talk like them!'

Catherine speaks freely, laughing, sharing the stories of the elders and her own confusion. In 1950 she went to the priest and said, 'Father, I am pained. I have the teachings of my ancestors and I have the teachings of Jesus and I don't know which is true.'

'They are both true,' he told her, 'and you must follow both.'

'After that I felt sooo much better, so relieved!' she says and laughs again.

It is 1986. Catherine is speaking to the camera as part of a television documentary based on Nelson's book.[25] She took Nelson into her home in spring 1975 because she believed in sharing the wisdom of the elders that she had been given and because she had the English to communicate it. Born in 1927 in Cutoff, the village replaced by Huslia following floods, Catherine learned English when she was a teenager by reading the labels on canned goods and other settler products and became adept at mediating between the agencies of the colonial power, the US federal government, and her people, the Koyukon.

After Nelson left, she continued in the same spirit, speaking and sharing her knowledge and memories on the radio. In one episode Catherine describes going with a shotgun, aged ten, to 'a good geese place', Willow Lake, in 1937, and the sky being 'black' with birds for twenty-four hours, night and day. 'That's how much bird there was. But now,' in the 1990s, 'there's basically no birds by comparison.' Even in the 1970s Nelson wrote of elders complaining about how quiet the dawns were, with so much less birdsong, and of the water in lakes drying up that used to be full of wildfowl.

In another episode she tells of fishing with her grandmother in the 1930s and being instructed in how to forecast weather by the spots on the nose of a fish: white spots mean cold weather is coming. Or how a single seagull far inland portends a bad fish year, while a whole flock signals a good one. 'Each animal knows way more than you do.'

They ate little fish in those days because they had to make their own nets, and cotton yarn didn't last, and nets of bark twine or sinew took a long time. Her voice is soft and low, patiently explaining the intricate web of rituals, taboos and customs: the routines of fish camp, the proper way to butcher a beaver or pluck a goose, how killed animals should be allowed to rest overnight to let the spirit leave the body and a bear's feet are cut off to stop its spirit wandering around. Each animal is reserved for different people to eat, since the characteristics of the animal are contagious. Only elders, for example, should eat the loon (a waterfowl called a diver in Europe, like the black-throated divers of Loch Maree) since consumers of its meat risk becoming clumsy like the bird, and elders are already so. This holds for other entities too: water should be used sparingly, lest the greedy drinker become heavy like water itself.

The form and the pattern of the natural world shape the imaginative landscape and the social relations of the humans that live within it. To go against the natural world becomes unthinkable, a kind of heresy, and in the Koyukon tradition this carries a price. The television documentary features a funeral for a boy who has taken his own life. Mothers are weeping around a fire and burning offerings, a huge white spruce cross is being erected on a fresh grave by his peers, and Catherine is shaking her head.

'Too many accidents,' she says. 'It's because we are not respecting *hutlanee*, the traditional rituals and taboos of the forest.' For the television audience, she puts it another way: 'The Supreme Court should make a law requiring people to approach the earth as they would another human.'

The last record I can find of Catherine is a radio programme produced by students from the Jimmy Huntington High School in Huslia in association with the Worldwide Fund for Nature. It is 2005. Her hair is white and she is wearing glasses and a cardigan and speaking more slowly.[26]

'They shouldn't be messing with the moon, you know. Our

elders said that something was going to change when you put a man on the moon. The moon is connected to the weather. And see what happened?'

The other elders sitting in a circle nod and join in. The students have noted how the spruce trees are suffering from heat stress, and Rose Ambrose says, 'Weather is getting too old to control its own self. Koyukuk River, the water is above the bank, terrible, terrible.'

Virginia McCarthy agrees: 'This is not the same land we grew up in.'

Marie Yaska says, 'All the birds have songs for us. The one song that we notice that really changed is the robin. It just sing half its song and then it go "Ha ha ha." I wonder why?'

I want to know how the Koyukon are faring now. I phone the mayor of Huslia, Carl Burgett. As we speak one sunny morning in June, the world described by Nelson and Catherine springs into life in my mind.

'We're pretty much an untouched people,' says the voice on the end of the phone with a friendly laugh. 'There's no other village for 200 miles upriver or 200 miles downriver. Only the Eskimos live further north.'

Carl tells me he is standing outside the cabin he is building out of spruce logs. I have seen a photo of Huslia, the irregular cluster of solid wood cabins with outbuildings and elevated caches behind them sitting among scattered stands of birch trees set back from the wide bend of the Koyukuk River, which stretches away to the west. In the distance the plain breaks up into irregular patterns of muskeg and forest until the flats meet the dark line of the Brooks Range to the north, the whole sliced by the huge sweeps of the glittering river. In the Koyukon language the name of the place is *Tsaatiyhdinaadakk'onh dinh* – 'where the forest burned the hill to the river'. Huslia is a version of the native name of the stream, the Huslee.

On this June morning the snow is almost melted and the last of the ice has gone from the river, says Carl. 'Green up' – the budding of the trees – and 'break up'– of the ice – are past, and night will not be seen again till August at least. Carl is in a good mood.

'It's a good time to call, yeah, this is a good time,' he says, laughing again. Like most Koyukon people, he doesn't keep time by a clock. He eats when he's hungry and sleeps when he's tired. And during the polar summer timekeeping can fall by the wayside. Children play outside at 3 a.m., and people stay up visiting and go home to sleep when the conversation is finished. It's just before noon in Huslia, and Carl is going out to the edge of the village to chop wood. For forty miles in one direction is *ts'ibaa t'aal* (black spruce) and for forty miles in the other is *ts'ibaa* (white). Summer is the time for logging, preparing for winter. In the deep cold, heat is more critical to human survival than food. A well-stacked wood store is a sign of prestige.

More than seeing Huslia, now I can *hear* it. Nelson's book is full of voices like Carl's, with a humble, open, melodious grammar that seems as much a part of the natural soundscape of the land as the wind singing through the spruce or rattling the birch, the songs of birds, the splash of salmon in the river or the chop of a paddle on the lake. Nelson quotes one woman, possibly Catherine Attla: 'Some people will hunt the loon, but me, I don't like to kill it. I like to listen to it all I can and pick up the words it knows.'

Of course, the rip of chainsaws, snowmobiles, outboards and propellers is an established part of life in Huslia these days. Carl tells me with pride that the village is growing: 350 people now call it home and a hundred students are in the school. But without roads and within the protected enclave of the Gates of the Arctic National Park, where commercial logging is forbidden, the rest of the world has paid the Koyukon little attention. It's only recently that the community washeteria (permafrost means there is no running water) has been equipped with wireless internet.

'It's a fly-in,' says Carl, 'and that means isolation, traditional life, wild food.' The Covid pandemic has interrupted flights, but more than half of the community's diet comes from the land, so if the planes did stop, they'd manage. 'Our knowledge, our culture is still strong,' he says.

I ask him if Nelson's book has served the purpose the author hoped for, providing a reference for the Koyukon themselves in keeping their culture alive.

'What book?' he says. I remind him of Nelson and Attla.

'Ah, that guy.' He chuckles. 'Folks don't usually open up to out-siders, but that guy, somehow he pulled it off.'

Carl avoids my question about climate change and points instead to the terrific rain they've been having. The far north of Alaska often saw less than fifteen inches of rain in a year, but this year is a big rain year. And a big pollen year. The rain has been sweeping down over Huslia from the mountains steadily for months. Three times more rain than usual. The trees are loving it. As we speak, the sun is warming the resinous gum of the blonde-brown seed cones of the white spruce. As the resin gives way, the spring-loaded scales of the cones snap open in synchronised release. Looking down on a valley of spruce during the days of cone release is like overlooking a distant battle. The forest echoes with cracks like gunshots and clouds of yellow dust drift on the thermals above the canopy. On all the standing water, the lakes around Huslia and the surface of the river, are long yellow streaks of pollen.

People are happy, says Carl. For ten years the lakes have been drying out and forest fires have been out of control. 'It'll be a good year for berries – the forest should flourish this year,' he says hopefully.

The extra water means something else. The dynamic meanders of the river will take even bigger bites out of the riverbank than usual and more houses will need to be moved. The community moves four or five houses every spring but sometimes they're not fast enough.

'Yeah, that's normal,' says Carl, 'these days.'

This, maybe, is how we will all adapt in the end. The elders of Huslia are no longer complaining of the unnatural flooding, like Virginia McCarthy and Rose Ambrose in the radio interviews a decade earlier. Instead, the unnatural has become natural, the apocalyptic has become prosaic, a perennial event merging into the background. This, perhaps, is one of the emerging realities of climate breakdown: that grief is a luxury. The urgent demands of daily life allow for no such respite or detachment. There is always work to be done.

'The negative impact of climate change is on nature, not on people, because we adapt. If one species suffers then another flourishes,' says Carl matter-of-factly.

Carl is less worried about the gradual ecological changes wrought by the oil that for fifteen years he helped to extract while working on the North Slope than he is by the sudden ones planned by humans. Under the terms of ANCSA, all land and mining rights were vested in a dozen or so indigenous corporations. The Koyukon do not have any oil concessions but they, along with all indigenous groups in Alaska, owe the material improvements of the last forty years to oil, and that leads to a reluctance to criticise the industry.

As Roman put it, 'Sure, there's change. There are cherry trees in Fairbanks. But everyone has $500 a month in their pockets. We haven't paid state income tax for thirty years and there's zero unemployment. It still looks pretty nice.' It is the same bind in which all modern societies are caught, although the role of fossil fuels in the Alaskan economy is more explicit. Alaskans know better than most that action on climate change means major changes to comfortable ways of life. For many that is unacceptable. The hydrocarbon compromise, as Roman terms it, is still largely intact.

The previous autumn, in October 2019, Nanieezh Peter, fifteen, and Quannah Chasing Horse Potts, seventeen, two teenage girls, challenged the elders at the Alaska Federation of Natives Convention to force through a resolution declaring a state of emergency on climate change.[27] But the majority of the indigenous corporations are still pro oil. Two weeks later the Bureau of Land Management proceeded with the auction of four million acres of drilling rights on the North Slope while Conoco Phillips announced a $5 billion project called Willow. Along with heat and beavers, willow is what will soon dominate the melting tundra where the company now refrigerates the permafrost in order to stop its ice-engineered infrastructure from collapsing.

Rather than the accelerating fraction of carbon dioxide in the atmosphere, Carl's anxiety is instead focused on a more visible, less insidious foe: the US government. He is worried that the federal government will use the pandemic lockdown to ram through unpopular plans for the construction of a 216-mile road through the Gates of the Arctic National Park connecting the Dalton Highway to a controversial mining concession on the Kobuk River near Ambler. A road would reduce the Koyukon's ability to control visitors. But more than that, the proposed road would mess with the hydrology of the catchment, altering the drainage and changing the structure of the forest and the ecosystems within it. Over the last two years Seth, Roman, Paddy, Carl and 100,000 other people have signed a petition against the Ambler Road. At packed public meetings in Anchorage people angrily shouted down the proposals, but when we spoke in 2020 the Trump administration was rushing the approvals process along.

'I'd like those road-building guys to come down here and see what it is they are proposing to destroy,' says Carl. He isn't laughing any more.

What they are proposing would destroy the very foundations of the Koyukon world. Nelson recorded one of the Koyukon origin stories:

> In the distant time, Raven killed a whale in a lake and strung its innards around the shores. Since then, spruce trees growing along the lake-sides often have long, skinny roots. Mink-man went to tree-women and told them their husband Raven had been killed. When one woman heard the story, she cried and pinched her skin. Then she was changed into the spruce tree with its rough and pinched skin. When another heard it, she cried and slit her skin with a knife. She became a poplar with its deeply cut bark. When a third woman was told the story, she cried and pinched herself until she bled. She turned into the alder, whose bark is used to make red dye.

The land is the basis of the Koyukon world view. It is not just a larder, but a dictionary and a bible, the repository of stories, history and culture, something so sacred that it can never be replaced. Every spot and every species has a role in the story of the world that Raven made. A mine is not part of the story.

The mining project has undergone three name changes so far and, like a hydra, it continues to morph to stay alive, even though the projected tax revenues from the mine that the road would serve are dwarfed by the $500 million cost of the road to get the copper, zinc and gold out. By contrast, before cutting a spruce down later that morning, Carl will address the tree, give thanks and explain why he is doing so, because in the Koyukon tradition trees can never be cut without a reason as they give so much.

As well as the practical gifts of heat and shelter, the trees are respected for their medicine. The great old spruces of the forest are regular camping sites and, the Koyukon believe, protect those who sleep beneath them. The actively dividing cells (meristems) at the very top of each tree are where the power and the medicine of the tree is collected; the spruce head is the shaman's brush, which is used to take away sickness. The threatening power of the lakes of *Hudo' Dinh* and *Hunoo' Dinh* near Huslia can be neutralised by carrying the top of a spruce sapling when crossing them.

Western science agrees. Between twenty-one and twenty-five medicinal biochemicals are found in the spruce. They are concentrated in the growing tips of the tree and in the resinous gum that forms a protective sheath over the emerging leaves. This gum is a cardiotonic that helps with the oxygenation of the blood, reduces blood pressure and helps regulate arrhythmia in the heart. The dispersing agents that fire these biochemicals into the atmosphere have antibiotic and antiseptic properties – the same chemicals that do the work when you use pine disinfectant in your home. In their billions, the meristems of the spruce trees of the north are actually sterilising the air we breathe.[28]

In a surprising symbiosis, the spruce multiples this antibiotic function by encouraging the lichens that live among its branches to do the same. The needles of the spruce release an alkaloid called ethanolamine which circles the crown of the tree and triggers the manufacture of antibiotics by lichen. These then hitch a ride on the other aerosols of the spruce and sweep down, carried by the winds pumped by the forest, to disinfect the airways of the northern hemisphere.[29] Among the cocktail of aerosols released is a sticking compound called betaphellandrene, which acts like a glue. The antibiotics released by the tree have their own adhesive which sticks to exposed skin from where they are absorbed into the bloodstream. All this is carried in the fragrance of the tree. No wonder Japanese forest bathing in coniferous forests has been shown to have positive effects on health and respiration.[30]

The spruce is not doing this work just for humans. The bio-chemicals in the tree also attract insects that use its gum to build and disinfect their homes and its pollen for protein to build their bodies. Pollen is high in essential amino acids that insects need, and the honeydew produced by insect damage to the tree provides easily accessible soluble sugars for the insect world. The insects provide food for creatures further up the chain, in particular for the flocks of migrating birds that come to the boreal to breed. The birds in turn groom the tree of insects.

This is the moral order to which the Koyukon belong, in which all living things have a place, a voice and a soul, and with whom humans have individual relationships. The order that stands in the way of the Ambler Road. Nelson quotes an elder,

'The whole of Alaska is just like something inside the palm of a porcupine's hand.'

On 23 July 2020, while the pandemic ravaged America and com-munities across Alaska were locked down in their homes, the US Bureau of Land Management issued a right of way permit, approv-ing the route of the road through federal-protected lands. Nine environmental groups sued in the District Court of Alaska, argu-ing that the bureau had violated the Clean Water Act, the National Environmental Policy Act and the Alaska National Interest Lands Conservation Act by pursuing the project.[31]

While the Biden administration reviewed oil exploration in Alaska, it still pressed ahead with the Ambler mining project and the controversial road. And so the last hope of the Koyukon and those opposing the road is that nature will intervene and the collapsing permafrost will make the whole project unaffordable. The US government is, finally, taking some steps towards acknowledging the problem of carbon emissions but it has yet to demonstrate an understanding that the crisis in nature is not about global heating alone. It has heard the voices of oil, mining and finance executives but not the voices transcribed by Richard Nelson; the voices echoing out of the far reaches of the cold northern woods with the grammar of another world: 'The country knows. If you do wrong things to it, it feels what's happening to it. I guess everything is connected together somehow, under the ground.'

5. The Forest in the Sea

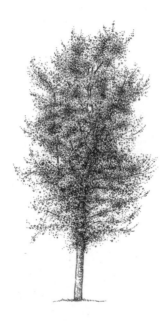

Balsam poplar, *Populus balsamifera*

Merrickville, Ontario, Canada
44° 55' 06" N

From the Brooks Range in Alaska the treeline cuts straight across into Canada's Yukon, the leading edge of white spruce following the contours of the most northerly parts of the Northwest Territories. It appears to be heading for Nunavut – the province of Baffin Island and Canada's Arctic archipelago – but it never gets there. Instead, the trees stumble dramatically south in a great green cascade down through Alberta and Manitoba to meet the sea at Hudson Bay at a town called Churchill.

Churchill is located at fifty-nine degrees north, the same latitude as John O'Groats, the northern tip of Scotland, but its mean annual temperature is a great deal colder. The daily average minimum in January is minus thirty. In winter Hudson Bay freezes almost completely, up to two metres thick, and is not entirely free of sea ice until August before freezing starts again in November. Hudson Bay is the world's largest bay, cycling water from the Pacific and Arctic Oceans, mixing it with one third of Canada's freshwater drainage and sending it out again through the Hudson Strait down the coast of Labrador and Greenland into the Atlantic. It is this long icy reach of the Arctic Ocean into the heart of the North American continent that pulls the temperature down and the treeline with it.

At Churchill three ecosystems converge – the sea, the tundra and the trees. This three-sided ecotone has led to the town's reputation as the polar bear capital of the world. Bears leave the rotting sea ice in the summer to forage on land and to den on the tundra and in the forest, giving birth in the autumn before returning to the ice when it freezes.

Here, I reason, is a key point on the treeline, an ideal spot to come and observe the changes afoot.

'No, no, no,' says the soft Irish brogue on the other end of the phone. 'You can't just look from one edge. You can't understand the relationship between the trees and the ocean without appreciating the whole watershed. You've got to go upstream.'

Six months before the pandemic struck, in the summer of 2019, I made a pilgrimage to Canada, to visit one of the foremost scholars of the boreal forest. When I finally meet her, Diana Beresford-Kroeger is worried. The summer is too hot. It hasn't rained in Ontario for months. The trees in her garden are suffering. Her vegetables at least she can water, yet outside the car window on the drive from the airport, rows and rows of regimented corn stalks stand to attention in wilting green jackets as far as the eye can see. When she first moved here forty years ago, many of these oceanic fields were forested.

On the opposite side of the road a huge industrial crop sprayer rumbles towards us on wheels as high as a house, its long bouncing arms folded behind it dripping poisonous pesticide all over the tarmac. Diana recoils so strongly that I half-expect her to hurl her body out of the window.

'Damnation!' she spits. 'Pesticides. That's all the planet needs right now is more cancer. It's death by design for plants, animals and us.'

The sun is getting low when we arrive in Merrickville, a quaint colonial town with stone shops, a bulging block house and sleepy cross-hatch streets at the junction of the Rideau River and Canal. The site of an old sawmill, wool mill and grain mill, it was the end of the road for virgin logs floated down the river, connecting the route for timber destined for Quebec and Montreal.

Half of Canada is forest. Most of the other half used to be, but the southern sweep of the boreal band bordered hundreds of miles to the north by Hudson Bay has been progressively gobbled up by agriculture, industry and the cities and suburbs of eastern Canada. The current rate is 1 per cent deforestation a year. The 160 acres that Diana and her husband Christian have looked after since the 1970s is one of the few pieces of old-growth cedar forest left.

As we turn into her drive, the sun is at the height of the crowns of the red pine and spruce that line one side of the road. Black cherry and white oak line the other. The sunset is a beautiful crimson haze, blurring the outlines of the trees, but Diana doesn't like it.

'Particulate pollution has increased one hundred per cent this year. One hundred per cent! Did you know that? It's because the trees are going.' She explains that all trees, deciduous trees like maples especially, have trichomal hairs on the underside of their leaves that comb particles out of the air, which then get flushed to the ground when it rains.

Diana Beresford-Kroeger is one of those rare people who can alter the way you see. She has changed how many people, including forestry professionals and leading academics around the world, see a forest. In Richard Powers' novel *The Overstory* there is a character partly based on her life and work called Patricia Westerford, who conducts groundbreaking research into how trees communicate with chemistry using aerosols – organic compounds released like tiny kites into the air – and through their root networks and internal fungal relationships. The fictional Patricia is a world-renowned scholar whose work has inspired whole generations of subsequent research and whose books are bestsellers. The real Diana's research has indeed changed the way trees are seen and studied, but she has not had the recognition and success that one might expect. The reasons for this are entwined with the foundation of her unique scientific perspective. She refused a university professorship because she felt she could do more outside the institution to push awareness and solutions on climate change. She calls herself a 'renegade' scientist, someone who thinks outside boxes and connects unnoticed dots.

Diana's upbringing in post-war Ireland was far from conventional. When both her parents died when she was eight, she was on the verge of being sent to Cork's infamous Magdalene Laundry at

Sunday's Well. However, the judge, learning of Diana's aristocratic connections in England as well as Ireland, suggested she live with her uncle Patrick. A distracted polymath who forgot meals, he fed her mind from his considerable library. One fireside discussion lingered in her memory. It was about temperature. If the mean average global temperature were to shift upward by one degree Celsius, it would cause famine. Crops have evolved to ripen within narrow bands of temperature: too hot and cool-temperate crops will wilt, too cool and tropical crops will perish.

Diana's summers were spent in West Cork, in the small valley of Lisheens near Bantry Bay, where knowledge of the ancient Celtic world was preserved. Brehon law (ancient native Irish law) states that 'an orphan is everybody's child'. Diana was tutored as a Brehon ward: the sacred knowledge of the Celtic triad of body, mind and soul ending with the laws of the trees was to be hers to share with the world in time. She was told that she would be their last ward tutored in the old laws; there would be no more after her. She carried the responsibility as a sacred trust.

Diana learned that the first dew on the clover is sacred for the young women of Lisheens. And she learned many other rituals whose underlying biochemistry she would later demonstrate in the lab. She had the freedom and the confidence to ask big questions. When she first came across the photosynthetic reaction, she realised it was the reverse of breathing. The same elements and chemicals linked plants and animals in a mirror: carbon dioxide and oxygen. She learned too that nearly all the trees and forests of Ireland had been destroyed.

She wondered, *What would happen if plants – let's say forests – were removed from the planet? The answer is obvious. Life would be extinguished.* A reverse of the famous bell-jar experiment.

For her master's thesis at University College Cork in 1965 she studied how plants would react to a warming planet. For her doctorate from Carleton University in Ottawa she compared the function of hormones in plants and humans. In humans the tryptophan-tryptamine pathways generate all the neurons in the brain. Diana proved that trees have these too and use them to generate all the same chemicals – a sucrose version of serotonin for example – that we have in our brains. Her work opened up the possibility that trees have the neural ability to listen, to think, to plan, to decide, perhaps in the cambium skin, the inner layer of the bark. Later she embarked on additional doctoral work which revolved around the oxygenation of the beating heart. When oxygen levels are too low, cardiac damage ensues. She manufactured a new non-typing blood to correct this situation using a process called haemo-dilution. Today this artificial blood is used in transplant therapy and in the delivery of medicinal biochemistry within the body to arrest cancers.[1] The leaf and the heart are the two most important organs for human life on earth, and Diana's life has been committed to understanding, and protecting, the relationship between the two.

Her ideas were controversial, seen as almost heretical by the scientific establishment in the 1960s, so when, in her next move, she examined the cell of a plant under an electron microscope that she had modified herself and discovered bioluminescence, a phenomenon in quantum physics that twenty-five years later would earn a team of three scientists a Nobel prize, the powers that be at her university in Canada refused to continue to fund her work.[2] Three suited men behind an austere wooden table politely told her, 'You should go home, get married and have children.'

Diana turned her back on mainstream science, bought the farm with her husband Christian, built a passive solar house that reflects microwaves and got lost in the woods, where she established her own research garden and microscopy. The biodiverse forest of Ontario was her solace and salvation. The pristine botanical wonders of Canada were overwhelming for a young woman from Ireland who – thanks to the highly effective British deforestation of her native country, which has left many Irish people ignorant of the ecological history of their own land – had never seen the anatomy and sheer dimensions of a virgin old-growth forest. She conducted experimental trials, collected rare and endangered species from all over North America and planted an arboretum. She forged relations with the First Nations indigenous peoples of Canada, whose plant knowledge and wisdom she respected immensely.

She combined what she knew from botany, organic chemistry and nuclear physics with what they told her, adding to Celtic wisdom, and published the results in two landmark peer-reviewed reference books, *Arboretum America* and *Arboretum Borealis*, which explain in detail the critical role that the northern forest plays in regulating the water, the air, the soil, climate and the feeding foundations of the oceans and which catalogue the huge untapped potential of trees to supply food and medicines for the modern world. Among the Mohawk and the Cree peoples she is known as the keeper of medicines. There is no greater authority on the boreal forest. She is a prophet for our times.

We have just had breakfast in the white pine home that she and Christian built.

Diana, in shorts and a yellow Australian T-shirt with SURF PATROL on the front, impatiently brushes unruly silver hairs out of her face. She jabs an urgent finger at the boiling kettle sending coils of steam up into the rafters. 'That!' She points. 'You see that!' I nod my head. 'That's the simple physics of global warming right there! It's a basic principle of science. The higher the temperature the faster the rate of reaction.'

She is trying to explain to me why all forests are threatened, especially the boreal. Warmer temperatures lead to more evaporation, which leads to more precipitation, but the rain does not stay on the ground. The heat accelerates the cycle of evaporation and condensation, with more of the water converted to atmospheric water vapour. This is why warming is so dangerous. Not that it will kill humans by making the planet too hot for human habitation today, but the acceleration of the water cycle will cause drought and excessive soil moisture that will stress the forests and trees and the root systems of all the plants needed to oxygenate the atmosphere.

Diana has been obsessed with climate change since 1963. 'I just don't want the people and children to suffer,' she says. 'Especially the children.'

By the time I visit her, humans have triggered an increase in the atmosphere of carbon dioxide up to 415 parts per million. And while humans can survive perfectly well at 1000 ppm, the heating effects on the planet of such a thick blanket trapping radiation is what she worries about – the impact of heat stress on the oxygenating capacity of the plant kingdom on land and in the sea. Huge die-offs of plankton or trees, the end of rainforests and the desertification of the tropics could see an acceleration in the already declining proportion of oxygen in the atmosphere. Trees cycle about half the oxygen in the atmosphere; photosynthesising algae in the ocean maintain the other half. Both are projected to function less well as the planet warms.

'The sick, the young and the unborn will go first,' she says: those whose hearts will fail at slightly lower concentrations of oxygen; babies, who need a lot of oxygen in the first few years of life to grow; and foetuses, who require super-oxygenated blood through the placenta.

'The human female has evolved to gestate for thirty-eight weeks, not forty or forty-two. A human foetus needs a certain amount of oxygen, and they won't get it in thirty-eight. I would expect to see problems of reproduction, fertility, miscarriages – that sort of thing. And we might see that fairly soon. If not in my lifetime, then almost certainly in yours.'

Diana is rattling round her kitchen, fetching books and drawing diagrams on scraps of paper to illustrate the chemical reactions and ecological processes she is talking about.

'People must understand the fundamental importance of trees. And stop the madness of cutting them down.'

'*Populus balsamifera*, that's the one for you,' she says. 'You must, must look at that.'

The boreal is the 'last forest', according to Diana. She says that the Amazon is probably done for, even if the wanton deforestation were stopped immediately; fires and drying will see to it within fifty years. Other tropical forests are severely degraded, especially in west Africa, Malaysia and Indonesia although, overall, global deforestation has slowed in recent years, the abandoning of

farmland in Russia and Europe balancing out the destruction elsewhere. The boreal is the biggest and most important intact biome
stretching over a wide temperature range, and thus has a good
chance of adapting.

The key species within the Canadian boreal, Diana says, the one
that anchors the ecosystem and its relation to other ecosystems, is
the balsam poplar. It is a sacred tree for the First Nations peoples
and yields a strong medicine, the strongest of all the northern tree
species. The medicines of the north are the most powerful because
it is extreme conditions, drought and cold, that cause the trees to
produce these chemicals for their own protection. The Cree people
call the balsam poplar the ugly tree because of its knobby bark and
bulky leaves, but these are the physiological features that make it
so valuable – a treasure trove of medicines – and have led to its
other nickname – the balm of Gilead. The deep cracks in its bark
trap rain and funnel water to the tree's roots. Its large plate-like
leaves with heart-shaped tips and a bright green waxy surface are
full of oils and resins. And its large sprawling limbs, which reach
out and spread in a seemingly random pattern, provide beneficent
shade for the complex life of the boreal's understorey. It might
seem ungainly, but it is the dependable nurse of the woods.

The chemicals are stored in the tree, in the leaves and bark,
under pressure, and Diana throws up her hands as she explains
how they are not well understood – how little we know. Dozens of
labs and teams of scientists could work on this one tree. However,
Diana has made a start. She has watched the spring sun warm the
buds the female balsam poplar produced the previous autumn,
before winter's deep freeze. As the resin that has held the bud tight
all winter begins to melt, it triggers the bud scales that protect the
leaves to expand. As the weather warms up, the sun heats the resin
molecules and they take off into the air as esters and terpenoids.
These are aerosols, and this flush of tons of oleoresin into the
atmosphere from millions of poplars of the boreal every spring
acts like a health shield for all life on the planet. The aerosols, she
has discovered, are expectorant, anti-inflammatory, antibacterial
and anti-fungal. But this is just the beginning. The oleoresins also

contain dihydrochalcones, other flavones and acids which are essential for human brain, liver and glandular development. They are the building blocks of the brain and form the brown fat in the body essential for humans to withstand cold by shivering. The shiver reflex metabolises fat as fuel. The tree that has learned to withstand cold helps us to survive in the same conditions.

Also present are the prostaglandin group, relatively new to science, which includes prostacyclin, a vaso-dilator that helps the primary function of the heart and opens and cleans the arteries, and other oxytocins that aid female fertility and reduce blood pressure. The Cree nations used the poplar's sap to treat diabetes, which makes perfect sense given that Diana found populin with glucosides present in the tree. It helps the stomach and digestive tract to slow gastric secretions and regulate the metabolism of fat breakdown.

The northern trees interest Diana in particular because they have evolved in the harshest conditions. They have learned lessons, contain hormones and have survival strategies that other plants will need to learn with climate change. And they contain chemicals that are essential for humans, if only science could find the time and resources to pay proper attention. Then, she says, perhaps we would think differently about how we harvest trees – timber may in fact be the least valuable use for the forest.

The balsam poplar creates these chemicals from the minerals that it mines deep underground. Unlike shallow-rooted conifers, it has a deep taproot that acts as a conduit between the benthic and permafrost layers in the soil. This draws up minerals and concentrates them in the leaves. Also unlike conifers, it loses its large leaves in winter. A single tree can produce up to five acres of leaves if you laid them out. This huge volume of leaf litter is what makes it a keystone species for the rest of the boreal and beyond. The soil around balsam poplars is very dark because it is rich in fulvic and humic acid, with large molecules related to melanin and melatonin, the chemicals that cause skin colours in the human body. Pigments carry trace minerals in the soil; they can attract and lock in metals from the rotting leaves, especially iron – the essential catalyst in organic growth. These acids leach into the soil and the water table and eventually find their way out to sea. In saltwater they are the catalytic trigger for the foundation of the ocean food chain.

The minerals in balsam poplar are concentrated because of the cold. Unusually for a deciduous tree it has evolved to thrive at northerly latitudes, perhaps in response to stress or successive bouts of climate change – ice ages and so on – moving north in warm periods and then, thousands of years later, finding itself stranded when the temperature dropped.

One consequence of its adaptability is the poplar's ability to reproduce itself vegetatively. Both male and female trees can send out tuberous roots laterally underground, often a good distance, which then sprout into new trees that are somatic clones of the original. The poplar can create a forest all by itself, linked by a net-work of roots underground that store nutrients and transport messages, food and carbon among all the trees in a constant exchange that research is beginning to reveal looks an awful lot like mass computation. What appear to be young trees are often just sprigs of a much larger and older creature – especially if they are present in numbers. Stands of balsam poplar and its more dainty cousin, the trembling aspen, *Populus tremuloides*, are usu-ally markers of ancient forest across the northern hemisphere. The oldest living organism yet discovered is a stand of aspen trees in Utah, all connected over eighty acres, each clone bearing the same DNA of one ancestor, dating to the Pleistocene ice sheet melt, 1.6 million years ago.[3]

But this is just one of the three forests that the balsam poplar sustains. The second forest is the understorey – the shrubs, mostly berry-bearing bushes crucial to birds, mammals and humans of the boreal, which need the shade the poplar provides just as much as the minerals it excavates to survive. The third is the forest in the sea.

Years ago, when Diana was collecting specimens of seaweed from the shore in Cork, she wondered why the estuary was so rich with diverse life – seabirds, whales, seals and more. Was it just that freshwater held more oxygen than the sea? Or was something else going on? By chance she subsequently met Professor Katsuhiko Matsunaga, a Japanese scientist from the island of Hokkaido, who had asked the same question and come up with an answer.

Matsunaga was intrigued by the apparent connection between the clear-felling of a forest for agriculture in a province of Hokkaido and the simultaneous collapse of the ocean food web all along the coast. His research proved that primary production in the ocean – the growth of tiny unicelled creatures called diatoms and desmids that make up the mass of phytoplankton, the base of the ocean food web – depends on nutrients and minerals which do not naturally occur in the ocean in a format that the phytoplankton can access. They need phosphorus, nitrogen and iron to photosynthesise and divide, and they get these chemicals from trees.[4] Phytoplankton live in huge columns in the sea. Sunlight activates their chloroplastic structures, and they use the photons from the sun's gamma rays to split the carbon from the oxygen in carbon dioxide. It is a mysterious process; we still don't quite know how photosynthesis works, but the photons must be concentrated or magnified like a laser, so that they can split a tightly bonded molecule. Then at night the phytoplankton turn that carbon into proteins using nitrogenous enzymes and iron. Iron is the key, scarce, resource; it is the catalyst. Without it plankton cannot divide. The hydroxyl groups of the humic acid made by decaying deciduous leaves mean it can attach to things: to water molecules, to the bottom of the sea ice for example, hence the green fuzz on the bottom of icebergs – algae harvesting the acid. Without the cradle of the humic acid molecule, the heavy iron atom would sink straight to the bottom of the sea or the river; indeed it would never have arrived in the water in the first place. The humic acid carries the iron to the sea, where it is used by the plankton to grow.

When protein is available, phytoplankton use iron to reproduce and divide. Zooplankton eat the phytoplankton. Crustaceans, minnows, molluscs and mites eat the zooplankton. Fish eat them, and bigger fish . . . and so on. Iron made available by trees is the foundation of the food web in the ocean.

Famine on the land causes famine in the sea, as well as a huge reduction in the amount of oxygen. Blue-green algae accounts for over 50 per cent of all planetary photosynthesis, a vital source of oxygen. Drought is not the only problem; a flood too can be deadly. A flush of nutrients from the land – too many nitrates and phosphates from agricultural run-off – can create an anoxic zone in the ocean. Algal blooms – the overproduction of phytoplankton beyond the capacity of the higher food chain to consume – encourage outbreaks of bacteria that eat the algae but in doing so use up all the oxygen in the ocean, creating dead zones. When fish swim through a dead zone, they die.

It turns out there is something behind the old Japanese proverb, 'If you want to catch a fish, plant a tree.' Trees regulate the atmosphere through their own activity and through their stewardship of primary production in the ocean. And yet, Diana says incredulously, 'Can you believe that no one has characterised the humic acid molecule to date?'

The balsam poplar is the mineral miner par excellence and the largest deciduous tree of the boreal forest by far. It is not habitually a treeline species but it can survive at minus sixty-seven Celsius and can weather extreme heat. 'In terms of climatic range and production of organic material, none of the evergreens can match it,' says Diana.

To understand what is happening in the ocean in Churchill, where the trees, the tundra and the forest meet, she recommends I first visit a place called Poplar River, a First Nations reservation that forms part of the Hudson Bay catchment flowing into the Nelson and Churchill Rivers. It is one of the last intact watersheds, where one can observe the unimpeded functions of nature.

'The people there will be able to tell you all about its sacred significance, the medicines and many other things – more than I can know. But you'll need to be prepared for a different mindset. They don't see things the way you're used to, if you know what I mean.'

I explain that I spent many years in Africa living with indigenous people, studying the Swahili language and experiencing the spirit world, if that is what she means. She laughs and sends me on my way: 'Oh great, that's all right then. We're cooking with mustard here.'

Poplar River, Manitoba, Canada
53° 00' 07" N

The destination always starts at the terminal. You get a sense of the world at the end of the railway or the flight from the people lining up, the names of the stops, the content of the conversations. In Winnipeg the journey to the reservation starts in the shuttle bus to St Andrews Airport. I knew about the difficult history of the First Nations in Canada and the generalities of colonial expropriation, but I was unprepared for how present the struggles still are, how fresh the wounds and how high the feeling.

The driver is a guy called Murdoch, a Scottish name. He has a shaved head, glasses. He is proud of the fact that he is a self-made man and puts it down to his independence from a young age. He was taken away from his mother when he was ten. 'Best thing that ever happened to me . . . I made all my own money. Not like these Indians.' I take a slow breath. I presume I am getting the lecture because the only destinations served by St Andrews are the First Nations reservations to the north. Murdoch wants to make a point, although I am not sure what.

'The thing about these Indians,' he muses, 'they just want more money, more money, more money. How long are we gonna keep paying them? Yes, we stole their land a long time ago, but how long is it gonna go on? I can tell you it stops right here.' I am confused. The force of his expression is strange. Despite the Scottish name, Murdoch, it turns out, has a 'treaty number', which means he is the fourth generation descended from indigenous people and can thus claim housing, tax relief and the various other social benefits promised to Canada's First Nations peoples in the suite of treaties their leaders signed with the colonial and settler governments.

'It's absurd. I paid no tax on my new car! And if you say any-thing they call you racist.'

Murdoch's point becomes clear. He wants to be seen as inde-pendent, not a taker of handouts, but he is on unsteady ground. The history is unsettled and he is unsure of his relation to it. And so he risks becoming absurd himself: a mixed-race man repudiat-ing his own blood.

Outside, the view is of sterile squares of yellow, green and brown, a boxy light-industrial landscape of roads, suburbs, drive-throughs, prairie skies and barns walled with plastic siding. There is barely a tree in sight in what was once known as Elm City. This is what the white man, Murdoch's ancestors too, did with the land they stole. This is the 'civilisation' with which Murdoch is asking to be identified rather than the reservation or the majestic savan-nah that was here only a little over a century ago.

He is curious and wants to know why I am going to what he calls a 'godforsaken' reservation. Is it for a holiday? The conversa-tion turns to climate change and the death of Lake Winnipeg from agricultural pollution. Murdoch is pessimistic. 'There's nothing we can do about it. There's gonna be a nuclear war. Wipe out half the world's population and start again – that's what I think. Hope you don't think I'm too negative!' I move uncomfortably in my seat.

Two more passengers get into the minibus: a well-fed couple who manage supermarkets across northern Canada for the North-ern Company – what used to be called the Hudson Bay Company, a legacy of the old colonial networks.

Murdoch is on surer ground with the newcomers, who relax into the comfort of shared prejudices as they lecture me about what to expect.

'I hope you haven't got any whisky in that bag of yours – you'll be going to jail!'

They joke with Murdoch about the self-imposed regulations prohibiting alcohol on the reservations and the rife alcoholism and drug abuse that actually goes on. They moan about the quality of the food there, apparently unaware of the irony of running a food shop that might be in a position to improve things. They don't feel bad about the exorbitant prices charged to captive customers; First Nations people living in remote reservations often inaccessible by road face steep air fares to get anywhere else. They reason that the residents are spending federal welfare dollars and they don't pay taxes anyway. In a sense, their casual racism passed off as local knowledge is a necessary myopia in order for them to feel good about their job, to feel good about their role in a system of colonial profiteering still basically intact.

The airport appears to tell the same story of exploitation and abuse, although from the other side. Not in the language of justifications and irrationalities, but written in people: a proud culture struggling under the weight of alien consumerism, bodies bent out of shape by industrial food, crammed into synthetic clothes and hats promoting foreign baseball teams, speaking a language hemmed in and eroded by the yapping and barking of English, the same slow, wary gaze of diminished expectations, unkept promises and cynical hostility that I have come to know so well in many other former colonies.

By the time I am on board the twin-engined Cessna to Poplar River the colonial mindset is in danger of colouring my own gaze. The only other passengers are two broken-looking indigenous people, a man and a woman, slumped in their seats in stained clothes, snoring in an alcoholic haze before we have even taken off. Stripped of agency, in the white man's world, they are easily read as victims of a long arc of European violence in North America, to be pitied or pilloried depending on your political standpoint. But in one hour's time they will sober up, descend from the plane and resume their roles as parents, siblings, community members or elders on the reservation, custodians of ecosystems which benefit all Canadians and teachers of ancient wisdoms that we must all soon heed.

After twenty minutes, the tiny aeroplane crosses a border far below. The huge squares of agricultural land where nature has been engineered, trimmed, suppressed, poisoned and sprayed are suddenly gone. To the east, the muddy waters of Lake Winnipeg, the tenth largest freshwater lake in the world, lap a beach tinged green with algae. Below, the arrival of the forest is like a huge breath of fresh air. For the next hour random colours streak the earth – the speckled orange of peaty moss, uneven stripes of yellow aspen and the black of spruce, myriad green polygons of swamp grass, swirls of forest and threads of streams dotted with bright pearls of water on the muskeg: the Hudson Bay Lowlands, 300,000 square kilometres, the world's largest wetlands. They mark the beginning of Canada's fabled but often overlooked north – so much part of the identity of the nation and yet so little visited or examined.

It is tempting to be lost in the immense horizon, overwhelming for the eye as well as the mind: a pristine habitat without any human mark. But this is another colonial idea. No 'untouched wilderness' is either untouched or wild to those who call it home. A First Nations person in nature is never truly lost and always at home. The land below has been shaped by millennia of human guardians, the most recent of whom are my snoring fellow passengers.

It is a landscape to inspire humility, worship and service. How easy to imagine that it is limitless, how easy to pretend that it is invincible. The idea of wilderness, the excitement of it and its possibilities, depend on erasing its human occupants. The savannahs and forests celebrated by the early colonists were for the most part previously managed landscapes abandoned by native populations much reduced by prior waves of invaders.[5] Erasure is still ongoing, as the people – and their ability to use the law – are the principal barrier to capitalism's endless appetite for forest. And in this battle stereotypes – the degenerate native, pristine wilderness – are weapons. Both are necessary lies that underpin Canada's self-image of reasonable prosperity.

The website of Natural Resources Canada claims that the nation's deforestation rate is 0.4 per cent but conveniently assumes that all the clear-cutting in the boreal will eventually grow back and therefore does not count. In fact, once disturbed a precious ecosystem such as the one I am flying over, which has evolved over 30,000 years and is still evolving, can never be replaced. One seventh of Canada's boreal has been clear-cut since 1990, a shocking proportion going as pulp for toilet paper.[6] We are actually wiping our behinds with the last remaining trees that stand between life and death for humans on planet earth. Canada's forest disturbance rate (a better metric) is 3.6 per cent, the highest in the world, even higher than that of Brazil. The demand for pulp, paper and timber, as well as the vast tracts logged in order to reach the lucrative tar sands underneath Alberta, catapult Canada to the top of the table. The pressure and the incentives placed on the First Nations peoples whose signatures are required for logging and mining in their areas have been immense.

But for the last thirty minutes my plane has been overflying land that has been put beyond the reach of commercial corporations. This is one place where the seemingly inexorable process of environmental degradation for profit has been miraculously halted. Four indigenous communities led by the Poplar River First Nation came together to achieve a remarkable feat: in 2018 their traditional territory was designated a UNESCO World Heritage Site. The area of nearly 30,000 square kilometres under indigenous protection and management is the largest area of protected forest in North America, equivalent in size to the country of Denmark.

The UNESCO designation was not granted only for the environmental significance of the land but, in an important precedent, for the cultural relationship that the Anishinaabe people have with it. For the indigenous peoples who have lived here ever since the land rose out of the water in their creation myth, around 8000 years ago, the same as the Stone Age in Europe, do not imagine humans as separate from the land, but as part of a total system, one organism. Like all the other indigenous human inhabitants of the northern woods, they hold that the rocks, the water, the trees, animals, plants, wind, rain and thunder are all invested with spirits with whom they share the land and with whom they must negotiate for finite resources. It is this relationship that has been acknowledged and protected.

Below eventually appear a boat, a radio mast, a flash of iron roofing, clusters of buildings along the jaws of a river pushing clean water into the polluted Lake Winnipeg and the scar of an airstrip cleared in the forest. The one-time summer fishing camp of a particular band of Anishinaabe people nailed to an English map by the Hudson Bay Company in 1806 has become, by virtue of its natural harbour and the establishment of a trading post, a permanent settlement of 1400 people.

Abutting the fence of the runway, detached wooden houses sit in clearings surrounded by rough patches of grass. A dusty gravel road snakes away into the trees. At the gate a dozen battered-looking pickups coated in white dust are lined up awaiting the mail. Behind the vehicles, a single ancient building long since fallen out of use carries a painted sign in green, POPLAR RIVER, ELEVA-TION 760 FEET. My fellow passengers grab their bags, hug the people who have come to meet them and drive off. Like Huslia, Poplar River is a fly-in community. The road across the muskeg is only passable when frozen in winter, and even that is increasingly uncertain.

This tiny hardscrabble place nevertheless signposts the way to the future, showing us the only viable and realistic exit from the cul-de-sac of unavoidable climate change. The 4000 inhabitants of the largest protected landscape in North America are demonstrating how to reforge a harmonious relationship between humans and mother earth. They are trying to remind us of some basic truths. They call their traditional territory Pimachiowin Aki – 'the land that gives life'.

'If the land gets sick, we get sick,' says my host, Diana's friend, Sophia. It seems so obvious, so simple. How could we ever have forgotten?

One of the dusty pickups at the airstrip belongs to Ray and Sophia Rabliauskas, community leaders and activists, who have come to meet me. We rattle and shudder down the gravel track away from the airstrip through the town. The main buildings of Poplar River consist of cubes of differing sizes clad in white plastic. The largest is the supermarket run by Northern, then come the school, the community centre and the fire station. As we leave the centre behind, the trees close in. Thickets of trembling aspen fifteen metres high line the road – what the Anishinaabe call poplar in

English, *auhsuhday* in their own language. Every so often a gap in the trees reveals a home in a clearing. Through the leaves, the river flashes its silver skin, scattering light. The town, the houses in the shadow of looming trees, seem at the mercy of nature: cars, boats, barbecues left in yards are at risk of being swallowed by the forest in a couple of seasons.

A huge truck with music raging inside roars past and blasts dust through our open windows. We pass the neatly fenced compound of the Royal Canadian Mounted Police, next to it the 'group home' where foster children stay who have been taken away from their families and the office of the dreaded Children and Family Services. Towards the end of the demarcated reserve, where the road comes to a halt in a mixed grove of aspen, balsam poplar, birch and willow on a bend in the river, is Sophia and Ray's log cabin. Inside their beautiful home decorated with indigenous art, a grandson races a car along the living-room floor. The windows frame a magical view down the river of wild rice, swamp grass and the sun sparkling on water.

We talk about the CFS. It might seem odd but the roots of the effort to protect their land that led to Pimachiowin Aki are there, in the offices of the government, where the echo of the violence done to former generations is still playing out. The state that stole the land and then the children of native families, sending them to residential schools in the name of 'civilisation', is still removing children from their families, but now for their 'protection'. Unemployment, alcoholism, drug addiction and social deprivation are endemic among Canada's indigenous communities, with some of the highest rates in North America. The reservation bears the brunt since the free housing promised by the state is only available here. There are not enough houses. Many families are still waiting. Meanwhile jobs on the reservation are thin on the ground. Training opportunities are almost nil, and for high-school education the children have to go away to Winnipeg. Even plumbers and electricians are flown in. Lack of self-esteem and of meaningful employment can hit hard.

There is not a family on the reserve untouched by Children and

Family Services. Sophia shakes her head, her eyes downcast. With her long black hair, she doesn't look old enough to be a grand-mother. She and Ray tried for years to care for nieces and nephews in her own family who suffered from parents deemed neglectful or abusive by the Canadian state. More often than not they were denied permission to do so. Once a child enters the system, it's almost impossible to get them out.

Ray and Sophia asked themselves what was going on and made a connection. In the early days of colonial rule native children were encouraged to go to school. Then, as the mission schools expanded and the reach of the state became stronger, that encouragement turned into a requirement. But because many families moved about on the land or were far from the outposts of the government, chil-dren were required to go to boarding schools, often against their will. Ray and Sophia noticed that the children taken to these schools, who were then beaten, shaved, abused, had their mouths rinsed out with soap for speaking 'dirty Indian' and experienced other horrors, became troubled parents. The community discussed this and they and other elders proposed a path for healing: the victims of the residential schools and the later generations who had suffered from their legacy should return to the land, remember their traditional ceremonies and relearn their native language. When language and culture are derived from the land, the route to healing is to return to the land.

The church, the state and the school had all told Sophia's par-ents that the old ways were sinful, pagan, wrong, but Sophia's father had always told her stories in the evenings. He was wise and, more importantly, stubborn. He allowed his daughter to go to university but cautioned her: 'Unless you know who you are and where you come from, that knowledge will be useless.' Luckily, she remembered.

'We had forgotten to honour the spirits,' she told me. 'We are not separate from the land; we are part of it. The creator told us that. Even sitting by the river gives you healing, if you open your heart, your mind, to the land.'

The community started a programme of healing camps. They

chose a site one hundred miles upriver where families used to go hunting in the autumn and spring. They raised money to airlift teepees and supplies for long stays. Miles from any road or path, in the old days only accessible by canoe, it was a sacred site on the shores of what the white man called Weaver Lake but which the Anishinaabe are only now learning to call, once again, by the name their ancestors gave it: Pinesiwapikung Saagaigun – Thunder Lake in English.

Many people had not been there before. Sophia herself had not been there since she was a girl, with her father. The young people did not know the old ceremonies or the language or even their own traditional names.

'We do not use the word "lost",' says Sophia proudly. 'It's there – the knowledge – in the land. To be revealed to us if we listen and practise our traditional ceremonies.' The land is memory and archive. If you sit and watch the animals for long enough, you can come to know something of what they know. You can see the horses eating the leaves of the balsam poplar to ward off colic, for example. Or you can see the beaver eating the aspen to make his fur gleam. Sophia laughs about *Canada's Food Guide*, a handbook on nutrition produced by the government. 'It's totally inappropriate.' Traditionally, Anishinaabe didn't eat carbohydrates and dairy products. Many are lactose intolerant. But they were told, 'This is a healthy diet,' when in reality their own wild food kept them far healthier.

'When we eat the beaver, we are getting all the medicines that the beaver ate: the aspen, the willow, the birch, the water lilies.'

The healing camps were hugely successful. They ran for many years, with specialist ones for children with behavioural problems, people with diabetes – since chronic emotional pain goes hand in hand with diet – and camps for elders abused in the residential schools. Going back to their traditional ceremonies, foods and language gave the community the confidence to critique what they had been told about their own culture and to assert their rights and responsibilities in relation to the land. The process opened a door. It led to the campaign to establish the protected area of Pimachiowin Aki and a Goldman environmental award for Sophia.

Now the school runs a forest camp up there for the children, and Sophia teaches Anishinaabeg, the native language, in the primary school. When she started she asked the class, 'Who here is Anishinaabe?' and only two children put up their hands even though they all were. Now, when she asks the question all of them know what she is talking about.

Reconciling culture with modern life is not always straightforward. Aidan, their oldest grandson, comes into the room where we are speaking, flops on the sofa with an iPad and immediately starts playing a violent computer game.

'Can you at least turn off the blood?' Sophia pleads.

But the community of Poplar River has found a way to live. They have a touchstone to guide them: the land. As often as they can, Ray and Sophia and others take elders out onto the land to spark recollections, vocabulary and encourage storytelling.

The next day they are taking a boat trip down the river to Lake Winnipeg, and they invite me along.

I hold Abel's hand to steady him as he steps down from the orange fibreglass speedboat onto the smooth dome of rock. His black shoes hover near the waterline lapping at the island for a moment while he reclaims his stick from me but then he pilots his body up the slope before lowering it onto a bed of lichen. The small island is covered with the common scrub of the boreal – juniper, willowherb, Labrador tea and cattails. Stunted jack pines (*Pinus banksiana*) cling to the occasional crack in the rock. Soil is a luxury of older landscapes; the young Canadian shield formation has only seen vegetation for a few thousand years.

Abel's black eyes look at me then fix on another of the myriad little islands that make up the shoreline of Lake Winnipeg at the mouth of the Poplar River. It is as if the land has been shattered and the pieces scattered into the water. His smooth brown skin glows in the warm afternoon light, revealing pinpoints of grey stubble. Next to him sits his cousin. Albert is slightly older than Abel, but his hair is just as black and his mind just as sharp. He wears a bomber jacket and a baseball cap pulled low over his glasses.

'*Manitoo*,' says Albert slowly. 'That is the creator ... *Manitoo-pa*, that means "where the creator sits".' He indicates the lake all around us with an upward flick of his square chin. To see god in everything is not an idea unique to the Anishinaabe. Until recently, it was an axiom of most human societies. It is harder to chop down a forest or strip-mine a meadow if you believe god lives there.

The Ojibwe nations, of whom the Anishinaabe are one, believe that in the beginning the land was raised out of the water and given to man as a gift for his survival. In return, he was bound to care for it in trust. As the Laurentide ice sheet retreated 11,000 years ago, it revealed a melt-lake called Lake Aggasiz much larger than the present Lake Winnipeg. Some 10,000 years ago all the land of Pimachiowin Aki was still at the bottom of Lake Aggasiz, but a couple of thousand years later the land, no longer depressed by the huge weight of accumulated ice, had rebounded, and the rock upon which we are sitting emerged above the water. In the beginning the rock was mined by lichen, moss, jack pine and spruce. Then, around 5000 years ago, these early pioneers were joined by birch and poplar. The *mashkeek* (wetlands) of Pimachiowin Aki have remained relatively stable since, until now.

'The moose are going, the birds are going, the rabbits are going, the caribou are nearly gone.' Abel is downbeat. 'We don't see the young ones going out to hunt, to go trapping, fishing.' He believes the problem lies in not harvesting the animals provided by the creator. To harvest the animals is to honour them and their spirit. 'If we harvest them, they will come back.'

It is a controversial idea and unpopular with conservationists, but if hazel responds to coppicing by growing more vigorously and sweetgrass responds to plucking, why not animals too? The botanist and teacher Robin Wall Kimmerer has written of a trapper who, by selectively taking males, actually increased the population of marten in his area. It comes down to a fine-grain understanding of ecology, although Albert and Abel wouldn't use that word. They talk instead about sacred ceremonies and spirits.

'Before we kill, eat or pick something, we give thanks, and we give tobacco to the creator.' Abel lifts his stick and swings it in a wide arc. 'Everything you see is medicine. I know twenty-four kinds of medicine.' Abel is one of the medicine keepers of the Poplar River Anishinaabe community. One of the trees he was given was *muhnuhsuday*, balsam poplar, what the Anishinaabe call in English black poplar. It is a sacred tree, as all trees with vital gifts are, which is to say most of them.

'I never had sore teeth when I was a kid,' says Abel. 'My dad would pick me a twig of black poplar, and the teeth would just fall out. No pain.' He smiles in proof.

The heart-healing properties of the balsam poplar mentioned by Diana are known to the Anishinaabe elders: you cut a piece of bark at heart height the size of a human heart and boil it in water to release the cardiotonics. They have found that it can also work against cancer, an idea Diana supports. One of the many souvenirs pressed on me by Ray when I leave Poplar River is a jar of ointment made from balsam poplar buds that he says is a wonder salve for any skin condition.

White spruce is used for building, for tepee poles and for winter bedding, Abel explains. Dogwood is used for basket-making and medicines; aspen is used for snares; willows are used to make sweat lodges. The list is almost endless.

'We could be here all day and all evening, the things I know,' says Abel. Then he pauses. And abruptly changes the subject.

'The residential school, I was there four years. So many bad things happened. I was sexually abused by my teacher. Yes. I was raped. Every day we were beaten. My mum used to braid my long hair, she told me never to cut it, but when I got there, they cut it off. The school was in Cross Lake. I often think about that. Now I still have nightmares when I sleep. Sometimes I punch the wall. When I came home, who's my mum? Who is my dad? I didn't know who they were. I didn't know how to laugh or smile. I lost my tickle. I used to be ticklish. Not any more. Mum never defended me. She told me, "Don't bring anything in my house; take it back where you came from."'

We look out at the lake. Pelicans patrol a dome of rock far

away. Bald eagles circle pine trees on another island. Canada geese float serenely by as if they had all the time in the world.

For sixty years Abel never told a soul about what had happened at the residential school in Cross Lake. He was effectively abducted by missionaries when he was eight or nine. But since his experience at the healing camp, it's often the first thing he tells people.

'Yeah, he does that,' says Ray. 'Abel likes to talk these days. It's like he wants to get it *out*.'

The healing camps allowed Abel and the others who suffered to talk. And talking was the beginning of understanding the gravity of what had happened and what had been lost and devalued. So much energy had gone into trying to assimilate, to deny, to be taken seriously and to succeed on the white man's terms and in his towns. Talking was the beginning of learning to glimpse the possibility of justice, even if the way there was still unclear.

Ten thousand native children died in the residential schools. The Canadian state attempted to sever Abel and Albert and the others from their history, culture and language, in effect to change who they were. If who you are is one creature embedded in an ecosystem along with many others, then healing means restoring you to that inherited role, restoring your connection to the land, although the word 'land' is insufficient. *Pimachiowin* is everything, a system of a world. Even 'nature' is misleading, since it has come to mean something separate from the human sphere. *Pimachiowin* is a perfect demonstration of what anthropologist Eduardo Kohn calls an 'anthropology beyond the human' – a set of signs, symbols, relationships and meanings bigger than the human realm of consciousness, of which humans are but one part.

We are supposed to be looking for moose, to estimate the population so that a 'sustainable kill' number can be arrived at. Once a staple of life in Poplar River, moose are increasingly rare. No one quite knows why, but the changing structure of the forest, a thickening due to fire and warming, making it more inaccessible, seems to have something to do with the declining population. The moose are moving north.

The ageing fibreglass speedboat with white faux-leather seats is driven by Eddie Hudson, local councillor and board member of the Pimachiowin Aki Corporation. Eddie reclines in his seat, his plaid shirtsleeves rolled up, an elbow on the gunwale, his cowboy boots stretched out. The sun is on his tanned face and the wind in his greying widow's peak of hair. He is admiring the fading light kindling the fine sandy beaches of the shore orange and pink.

'When I saw trees, I saw wealth.' He studied economics. But the elders taught him to see his traditional home differently.

All eyes track the thick wall of forest on the mainland. Eddie pushes the boat into a shallow bay of swampy grass and jumbles of boulders to take a closer look at a new beaver dam that has appeared since he was last here. No moose.

The wind is chopping the surface of the brown lake into ever higher waves sugared with white. We tie up the boat at another shard of rock and gather juniper brush for a fire to make tea. Two sandhill cranes that were feeding on the far side go clattering into the pink sky making a noise very like their Anishinaabeg name: *oocheechuhg*. Just as we perch on the rocks to drink tea, a female black bear and cub tumble out of forest on the mainland for a swim. We watch them gambol along the beach.

'We nearly lost it!' says Eddie, seated like a satisfied king surveying his domain. 'We still might lose it! We sold our land to the crown!' Everyone laughs and looks at Albert. It was his great-great-uncle whose 'x' represented the community here on Treaty 5, which ceded these lands to Queen Victoria in 1875.

'But all the "x"s look the same,' he says in his defence. More laughter.

'At least now we have a say; we can protect it,' says Eddie. One of the conditions of the protected status was that the community had to come up with a 'land management plan'; they changed this to 'land-use plan'. 'They say the land needs to be managed. Nonsense. Let the forest manage itself, let the animals manage themselves.' Eddie is suddenly not laughing. 'We have to put things in the white man's language so he can understand it.'

The difference between management and use is important: the first implies dominion, the second suggests respect, permission and gratitude.

Along with the moose count there is a multi-year study to measure the drying of the peat bogs – the *mashkeek* wetlands that gave rise to the English word 'muskeg'. A carbon study commissioned by the community found that one acre of muskeg holds eighteen times as much carbon as a comparable acre of forest and 444 million tons of carbon are stored in the whole of Pimachiowin Aki. Carbon accounting could be the new economy of Poplar River.

The terms of the land use plan and the Anishinaabe's debt to their creator are the same. The land must be used. The spirits must be honoured. The sacred sites must be visited and guidance sought from the creator. Tobacco must be burned and fish caught. An expedition is planned to Thunder Lake for the weekend. It will also be a chance to part the younger generation from their phones.

It is getting late. The sun is nearly touching the ribbed brown surface of Lake Winnipeg, and the wind is chasing us home as we thread our way through the random boulders, but Abel and Albert are only just getting going. Albert is from the sturgeon clan and Abel is wolf clan, they explain.

'The animal is your ancestor, your protector. You must learn the story of that spirit, that animal and respect it,' says Albert.

He tells the story of how the hare got his long ears, about the hunter tricked by the wolf who turned into a woman, then he points at a sacred site on the shore.

'That was where our shaking tent was. It was our telephone. No children were allowed. We would feast all night and communicate with other tribes as far away as British Columbia, even Nunavut, before the church outlawed it.'

Even after we tie up the boat back at the pontoon, hoist ourselves into Ray's truck once again and deliver him to his home, Albert is still talking as if lives depended on his knowledge, which of course they do, including and especially his own.

'All the trees in the world are descended from one tree, a cedar, that is still living. The cedar is sacred, and the poplar, because the poplar sustains the river . . .'

We did not see any moose.

The next morning Sophia, Aidan and I stand in the sunshine where the road slopes gently downhill into the water and stops. This is the end of the short track through the reservation and the beginning of the wild. It is where for thousands of years canoes have been launched for journeys upriver. On either side of the river stems of wild rice sway in the current. The sun breaks over the surface of the trees, bathing the opposite shore in golden light that glitters off the surface of millions of tiny articulated aspen leaves. Ducks twitch in the shallows, a bald eagle takes off with a cry from a stump on the shore and banks above the forest on the far side, and on the surface of the water a fizzing mass of bugs swirls like smoke. Beyond, to the north-east, a wide mirrored stretch of river cleaving the forest marks the way ahead, the route of our pilgrimage to the source.

The smell is narcotic: juniper, mint, gentian, aspen, spruce. The air, I now know from Diana, is thick with pinene and other aerosols that cleanse the atmosphere and make each breath aseptic and medicinal.

Two canoes are being shunted into the water and outboards fixed to them. We are waiting for one more. Bags are thrown in, paddles, fuel, cool boxes, a fishing rod, an axe, a chainsaw, a rifle.

It's hot. Hotter than ever, Sophia says. Summer temperatures never used to get above twenty-five but now there's a heat wave every year. Canada as a whole is warming at twice the global average, and the north even faster. Sophia kneels to pray and offer tobacco at the base of the tree where the bald eagle had perched.

'He will follow us now, upriver, all the way. It is a good sign.'

The final canoe arrives and we are off. I am assigned to a sleek grey vessel that looks brand new. It belongs to Roger, who grips the tiller with authority.

'Mind the paint, Roger!' a voice shouts from the shore. 'That's six thousand dollars of canoe right there!' Once upon a time, dugouts were made from balsam poplar – because of its diameter and the fact that it cures without splitting. Nowadays new maple canoes are flown in from Winnipeg.

I sit on the middle plank. Clint, appointed by the Corporation as the land's guardian – like a park warden, is in the prow. I will stare at his back for one hundred miles. The three canoes cut wide furrows in the brown water. We motor in formation, saying little, the whine of the outboards making it hard to talk. The forest scans past, mighty and silent, drinking the people in. Balsam poplar is mixed with trembling aspen, spruce and jack pine. But the overwhelming species is aspen – it's one huge thicket. Beaver dams punctuate the shore and the inlets and creeks that drain into the Poplar River.

The wide sweep of the river is an endless, unchanging vista until we enter an eerie burned zone. The crisped orange needles of the spruce are still on their branches and the trunks are black. Thin poles of aspen remain standing here and there but mostly they have collapsed. The understorey is already rebounding with a vengeance: willowherb, fireweed, willows and poplar, the forest coming back, much thicker than before. The occasional mineral whiff of charcoal floats on the breeze. The roots and suckers of trembling aspen respond to fire with vigour. Like the balsam poplar it can reproduce through suckers, cloning itself. I would not be surprised if all of Pimachiowin Aki was one tree. Touch one trunk in Poplar River and it might register in a stem 1000 kilometres away on the other side of the reserve.

Eddie remembers that when he was a kid you could see through the forest and spot a moose in the trees. The aspen were at the edge of their northerly range and didn't crowd out the understorey. But now the composition of the forest is changing. The undergrowth is dense, hard for moose to move through. No more dappled shade from the aspens and the jack pine but dark thickets of shrub instead.

After an hour we come to the first rapids. The canoes must be emptied, then dragged over a ladder made of straight spruce poles up the bulge of rock beside the waterfall.

It is heavy work.

'Heave!' two people shout at slightly different times.

'Too many chiefs!' shouts Clint, and everyone falls about laughing.

The cool boxes, bags, fishing rods, chainsaw, axe and rifle are again loaded into the boats and we jump back in.

'Only twelve more rapids to go!' chirps Roger with a grin. It's going to be a long day.

In 1794 John Best from the Hudson Bay Company was sent south from the York Factory on the bay to explore the territory of Pimachiowin Aki. It took him three weeks and fifty-seven portages to reach Lake Winnipeg. This was enough to warn his colleagues off, and the Anishinaabe communities were spared the fur trade for another hundred years.

In the lead canoe are George, Errol and Aidan, Sophia's grandson. Aidan is loving it, the pushing and pulling, the joking of grown men. Errol is a young man from a troubled home, but out here he is a leader, leaping out at each rapids, pulling the canoes with the strength of two people and making everyone laugh all the while. George is the most experienced river man of the group. He pilots his fibreglass canoe with elegant skill, weaving through the shallows and charging straight up the minor rapids, surfing the stopper waves.

The middle canoe is captained by his cousin Guy, a steady older man whose dark glasses never come off for three days and who says little except to deliver the punchline to someone else's joke. Eddie and Sophia are his passengers. They are the elders of the group and sit out the portaging at the rapids.

At the third rapids the rest of us pile out onto the slabs of bedrock at the edge of the river, revealed by the low water level. Thin soil covered by pale green, almost ghostly lichen forms a crust on the geological shield formation beginning a dozen feet higher up. We haul the canoes out of the water then heave, curse, sweat and drag them over rolled logs placed on a track through the forest, where we are plunged abruptly into another world.

Gone are the dazzling light of the river, the wide vista of forest. Glades of lichen, their pale spiny fronds pricking the air like coral, swim in and out of view beyond the gnarled limbs of jack pine. Gangs of aspen shoot their straight grey boles skyward like gun barrels. Sunlight reaches the ground in ribbons and spots, or else in sliding beams where a fallen tree has torn a hole in the canopy, as if we are underwater. But this coral forest has air, and what air! Fragrant and heavy, like a perfumery. I realise, with a jolt, how rare this experience is, how few people have smelled and touched a 7000-year-old forest. So this is what 'old growth' means.

Perspectives open up and then shut down. Beneath the jack pines the forest floor is spongy with moss and lichen, like tightly woven fabric with the threads pulled taut. The understorey is thin. You could see a moose coming, or a bear. Below the aspens the soil is softer and black. Poplar and aspen soils have been shown to have much richer mycorrhizal lives than that beneath birch, for example, the other main deciduous pioneer, which is more akin to pine and spruce soils, its evergreen neighbours. The chemical environment leads to more mineral exchange, elevated soil pH and a greater soil moisture content. The larger leaves give more shade, which is better for berries.

This year's leaves have just begun to fall, too early, covering the thick understorey with crisped grey flakes. Blueberries, chokecherry, wild raspberry and many others jostle for space in the aspen's dappled shade, and at the end of the trail are the greatest treat: plump, velvety berries called saskatoons that improve eyesight in the dark. We lay down the canoe and snatch handfuls of the berries before the final heave. Then, on to another ladder-like structure cobbled together from spruce, and the canoes re-enter the water with a *whoosh*.

'Don't let go of the rope again, Clint!' Everyone laughs.

For hours the endless forest streams by. The outboards whine on. One hundred miles without a glimpse of another human being. Only the regular sight of our friendly bald eagle, appearing on a bend, perched on a snag, keeping an eye on our progress.

There are thirteen rapids around which we must portage, some higher than others. After the last one is passed by the light of the dying sun, the river suddenly widens and opens out into a broad lake with forested sides stretching away beneath a perfect magenta sky.

'Home free!' shouts Roger.

On the shore a huge dead pine tree marks the entrance to the lake, and atop its highest branch sits a lookout, a large bald eagle, his white head nodding as we motor past. Sophia sprinkles tobacco on the surface of the water.

Thunder Lake is enormous, almost a sea. The wind is dying as we finally reach an island close to the further shore and slide the canoes up onto a gently sloping dome of bedrock striated with the marks of hundreds of keels beached here over thousands of years. In a shallow pool at the edge of the lake I spot a tiny speckled brown frog. The boreal frog can survive all winter with 75 per cent of its body mass frozen solid, ice forming within its fatty cells, drawing all its defences back to maintain a small chamber around the heart. When spring comes, she emerges from her cryogenic state. I hope she likes the balmy water; it feels, and looks, like lukewarm tea.

Perched on the top of the island dome of rock is Guy's cabin. Antlers of moose cover the end wall. Outside, among the pines, lie generators, old canoes, aluminium sheets, propane tanks, a strimmer, paint pots, even a refrigerator. Inside is a bedroom piled high with mattresses and a kitchen with a table, a cast-iron stove and a more modern gas one. The sink empties into a bucket; the cupboards are crammed with a year's supply of canned food, and there are live bullets in the cutlery drawer. On the wall is a picture of a wolf and a clock stopped at 6.44.

The five Anishinaabe men sleep in the one bedroom. Sophia and her grandson, Eddie and I pitch our tents on the tiny island. We get the short straw. I manage to find a relatively level glade of pale green lichen between stunted jack pine trees that crunches and crackles like stale bread underfoot. As we turn in and say goodnight, there is a shiver in the air, the jack pines combing the wind. In the sky above the forest on the far shore is a smudge that seems darker than the rest. Then an unmistakable rumble.

'Thunderbirds!' says Eddie, smiling. 'They know we are here.'

*

Fire is the driver of life in the boreal. The landscape and the forest would look very different without fire, indeed the evolution of the species here would be different too. The three broadleaf trees of the region – the trembling aspen, the balsam poplar and the paper birch – are all well adapted to fire. The poplars (the aspen and balsam) may look fragile with their pale grey bark and whip-like shoots, but when fire consumes the forest and even the soil, their taproots survive. Three to four weeks later suckers will begin to appear – the white-grey shoots of the aspen and the pale greeny-red of the balsam poplar. They thrive in the mineral-rich soils revealed after a burn. So too do the seeds of the jack pine, which without fire would not germinate at all.

Jack pine cones are hard as stone, their scales stuck together with resin that has the force of superglue to protect the seeds from rodents. At fifty degrees Celsius the resin melts and actually fuels the fire, transforming the cone into the wick of a candle that will burn for ninety seconds, enough to cause the cone to open like a flower. So the jack pine actually regulates the fire, making sure its seeds get just the right amount of heat for the right amount of time. The cone will then only release its seeds when they are cool again, by which time the ground has been cleared of competition and the seeds can germinate in the sandy mineral soils that they love.

The response of these tree species to fire is critical in the cycle of life that the Anishinaabe depend on. In the early years after a fire, the opened-up forest is a haven for moose and hares, which like to eat the young shoots and smoked leaves. For twenty years or so, such places are regular feeding grounds for moose and attractive areas for trapping the marten and lynx that come in search of hares. After fifty years, the hares begin to decline as the jack pines grow beyond their reach. Owls nest in the burned trunks, and by this time feather moss has begun to carpet the ground, holding in more moisture for berries and shrubs. Aspen stop reproducing after sixty years or so and then the moose begin to decline. The moss gives way to lichen after seventy-five years, and caribou arrive to graze it. Balsam fir and other conifers then take root in the deepening black soil – known as *okataywikamik* by the

Anishinaabe – created by the deciduous leaf litter of the poplars. By now the forest is not such a good food area for humans, and the people pray for it to burn again.[7]

Fire is a creative life force, and balsam poplar was always the source of fire for the natives of the forest. It makes the best socket for the bow drill used to generate a spark through friction, and nothing works as well as the inner bark of balsam poplar for feeding the ember generated by the drill. Decayed poplar heartwood, both balsam and aspen, which is soft and punky and glows slowly, was used to carry fire. And sometimes humans would set fires when the lightning didn't come or to encourage regeneration in particular areas. The Anishinaabe don't do that any more, as no one knows where it might lead.

The forest is drier. The fires burn hotter and for longer. More of the peat and organic soil is burning, and the species that take advantage of the severely burned areas – the fireweed and willows – are thicker and more aggressive than the delicate understorey of berries and tea. The jack pine has been hit by budworm and the poplars dominate. It is the young jack pine that the hares love most of all. Poplar has trace minerals that it uses to make the leaves of very young trees taste bad to hares. What will the hares eat? And what will the carnivores eat? And . . .

No one has ever seen a thunderbird. They hide behind dark clouds, but they are represented in pictures as birds with lightning crackling from their eyes and fire pouring from their wings. Eighty-five per cent of fires are caused by lightning, the rest controlled or accidental burns caused by humans. Thunderbirds are therefore the most important beings in Anishinaabe cosmology. Thunder Lake is named for them. It is a place where thunderbirds nest.

They are noisy tonight. I wake at midnight to a terrible crack that makes the night shudder. The jack pines are thrashing against each other and the waves on the lake sitting up like frothing dogs, their teeth flashing in the light. The lightning makes the whole sky course and flicker as though the world is inside a strobe. I give thanks there are no balsam poplars on the island – the trunks of green balsam poplars with their high moisture content are prime lightning conductors. Instead, I worry about the metal poles of my tent, perched as we are on a forested dome of rock six metres above the lake. For the rest of the night I roll around snatching morsels of sleep while the wind tears at my tent and the terrible rolling, rumbling slowly dissipates to the north.

The morning breaks still and clear without a cloud. The air is moist and fresh, the perfume of the forest activated by the rain. The earth smells sweet and the ground lichen, which was stiff and brittle the night before, has soaked up all the water like a sponge. Moss that looked dead and brown is suddenly pulsing green. The thunderbirds have poured the forest a much-needed drink.

Later that morning on the other side of Thunder Lake our three canoes enter the neck of another river. Older pines and large poplars line the banks. All along one side is a wall of rock. Roger slows our canoe. The granite is streaked with ochre. The canoes queue next to the cliff beneath a perilous-looking overhang. In a little nook are the remains of candles, cigarettes, plastic objects, coins and twigs. Roger stands up, takes two cigarettes out of his pack and places them on the stone ledge. Clint does the same.

'Grandfather site,' says Roger formally, by way of explanation. Even rocks in Ojibwe culture have animate properties. And the ochre paintings are the work of *memegwesiwag*, semi-human cave dwellers who taught the Anishinaabe how to make arrows and pipes out of stone. Roger and Clint laugh when they explain these things, but I do not doubt their faith.

Further on, we beach the canoes at a wide pool beneath rapids. George and Errol grab fishing rods and cast spinners into the race. Within minutes they are reeling in white and green striped fish with a spiny dorsal fin – walleye or pickerel.

'We're going shopping!' says George. 'This is our supermarket.' And he makes it look as easy as tossing tins into a basket.

Clint and Eddie pull dead logs from the forest and make a fire using dried grass and birch bark as tinder. The Anishinaabe only ever make fire with dead logs; they never cut a living tree.

As the kettle boils, Eddie and I sit down beneath a scraggy-looking pine. Eddie twists one of the bolls off the tree and cuts it in half with his knife. Inside is a sticky orange goo containing five or six creamy white worms – larvae – about as long as an ant. Budworm. I take a closer look at the tree. It is twisted and deformed. The flying insects lay their eggs in the buds of the pine, and these feed on the sap to create a large boll, or nest. All the trees are infected. For the rest of the trip I am unable to spot a healthy jack pine. Every tree along the river has the crooked look of a stooped old person, needles falling prematurely.

'Are you worried about global warming?' I ask.

Eddie stretches out his cowboy boots and tells about an elder.

'He said that the climate was changing and that the people would need to adapt. Even though it is raining more, the extra water will not stop the land from drying. It is getting warmer and it is drying out. The peat is burning these days – it never used to. The species will change, new species will come. The fish are going to deeper places in the lake, cooler. Eventually they will die, species like the walleye will suffocate. Then new fish will come . . .' Eddie is quiet for a moment as though coming to terms with what he has said. Then he shrugs and crosses his legs.

'I'm not worried. We will adapt, like they told us. I wouldn't mind a longer summer.' He chuckles.

Suddenly everyone has stories of change, of snow becoming heavier and wetter, of the ice roads melting, of the texture and colour of the ice on the lake, species like magpies and vultures appearing, martens and skunk, of berries tasting different, the picking season getting shorter, not enough snow for trapping in the winter, the increase in forest fires, the water level of the lake going down, the ancient scratch marks of canoes two metres higher on the rocks.

'Look at the leaves of the aspen,' says George. 'They are scorched.' It is true. On nearly every tree leaves that should still be green and healthy are singed with orange two months early, the stain of heat stress as sap retreats from unviable leaves.

'And the spruce?' I ask, noticing for the first time all the light

brown crowns of the Christmas trees among the stands of
forest. They seem to be drying too.

'Yup,' says Guy.

My gaze shifts. The pristine environment simmering with life so
apparently untouched and uncorrupted by industrialisation suddenly appears brushed with the first hint of death. The trees are
innocent victims of processes far away. Pathogens like budworm
can wipe out whole ecosystems. But ecosystem collapse is usually
documented after the fact. A forest can mask collapse for quite
some time because the decline of the keystone species and natural
processes might only be visible over long timescales. Conservation
groups are already talking about the 'transformation' of the Canadian boreal, as the whole structure of the forest is 'reconfigured'
by warming, but 'collapse' might be a more appropriate word.[8]

Research on the 'de-megafaunication' of forests – the removal
of large ungulates like moose, caribou and bear as they move
north, and the implications for biodiversity – is only just beginning. If the jack pine goes and the aspen is limited by the heat, the
balsam poplar could outcompete all other species. There are
already some scientists who consider a coniferous northern forest
a thing of the past.

'Pine get sick every two years or so, it's a cycle. They will come
back,' says Eddie, rallying. The others nod, keen to believe him.

'How long have they been like this?' I ask.

'Five years now,' says Guy.

'We will be fine,' says Eddie. 'We will adapt.'

The future is not a safe place for the mind to linger.

George breaks the awkward silence by ostentatiously opening a
can of Diet Coke with a loud hiss.

'We live off the land!' he laughs, raising the can ironically, and
everyone laughs with him.

That night, after we have eaten and given thanks for fifteen walleye fish between nine of us, Roger takes the chainsaw to a rotted
jack pine stump and we have a bonfire. The reflected flames dance
on the surface of the silver lake, flat calm in the moonlight. Just as

everyone is getting ready to go to bed, a glimmer appears directly in front of us. A light show is beginning between the lake and the sky. Behind clouds shaped like eagles in flight, beams of light begin to pulse a faint greenish colour. The glow strengthens and slips the bonds of the clouds to spread all across the northern dome of sky. Green light flows like water over galactic pebbles, rippling over the black, and behind it all a kind of immanence makes itself felt.

The Anishinaabe say the aurora borealis has the sound of a rattle, and if you clap it will go away. Eddie tries that but it doesn't work. Sophia wakes her grandson to witness it, and we all stare at the shifting, pulsing show, transfixed. Something about the light and the moment moves my companions to speak in their native language. The frequency of laughter increases. Aidan cannot understand most of what his grandmother is saying but that is not the point; she wants him to experience this, to braid it into his memory of who he is and where he comes from. It is impossible to tear our eyes away from the animated presence, which only grows in strength. By one in the morning the aurora covers half the sky, its rhythmic pulsing mirrored in the lake.

'The spirits of our ancestors are still strong in this area,' says Sophia. 'I want my descendants to sit here and enjoy the same spot as I am doing today.' She knows climate change is coming, she knows it will affect the things she loves and cherishes the most, but her solace and solution is the land.

'When we get there, we will be able to survive from this land that was given to us,' says Sophia. I think she is right: the people of Pimachiowin Aki are among the best prepared on the planet for global warming. Even as species change and new ones come, their reservations are remote and inaccessible, and they are in control of 30,000 square kilometres of land that can surely feed, clothe and shelter them, if not in the manner that they are used to. The forest is a life raft. And the knowledge of their ancestors to guide them is, for now, alive and well.

But the point of Pimachiowin Aki is not as a resource for the Anishinaabe alone. 'We believe we are making a strong contribution to the rest of the planet,' says Sophia. Most indigenous teachings speak about balance, about fire and water and the sacred relationship between the two. This is the knowledge, the message, the sacred duty that Pimachiowin Aki represents.

The Anishinaabe speak of the Prophecy of the Seventh Fire. Their ancestors described the eras of the Anishinaabe civilisation in terms of fires. The first refers to their beginnings on the shores of the Atlantic, the second to their move west, the third speaks of their need to move where the 'food grows on the water' (the wild rice of Pimachiowin Aki), the fourth of the coming of the foreign people from the east (the European colonists), the fifth describes their near-destruction at the hands of black-robed people with black books (the missionaries) and the sixth when 'the cup of life would almost become the cup of grief'.[9] This is the time that has just passed.

The people of the seventh fire must make a choice. To use the fire stick they have been given as a creative force, to choose the path of healing and nature which does not lead forward but back, to relearn the sacred teachings of the ancestors, the spirits and the land. Or to press on towards oblivion. Only if they make the right choice will the people of the seventh fire be allowed to light the eighth. A fire of renewal in a new world, which will be different from the old. The Anishinaabe of Pimachiowin Aki and other intact indigenous cultures are the fire-keepers of our earth, not only custodians of their own traditions and values but of a way of seeing and acting respectfully in harmony with the living world.

Aidan has fallen asleep on the rock. It is time to go to bed. When the sun rises again on the pure copper waters of Thunder Lake, it will be time to trace the effect of the sacred work of Pimachiowin Aki, to follow the influence of the catchment – downstream.

The canoes bounce along on the waves, heading back west. Everyone is dipping their water bottles, filling them from the lake, which they believe to be purer than the river back home, although both are clean enough to drink. Eddie tells me that black spruce contains a chemical that purifies the air as well as the water, and that when the sap falls in the winter it does something to the water table that helps the fish survive under the ice. Another nugget of wisdom for science to investigate.

At the end of the lake, as we enter the headwaters of Poplar River, Clint lifts his nose. 'Smell that? The scent of home!' he shouts. Like a salmon Clint claims he can tell his home river from its water's unique chemical signature. When trees are cut down or pollution introduced, a river's signature changes or becomes scrambled beyond recognition and the salmon stop returning. For a river to hold its scent all the way to the ocean, or enough of it to be recognisable despite the industrial catastrophe that is Lake Winnipeg, is quite a feat. The acerbic notes of that rich black soil must be very durable. Hundreds of miles downstream in the icy waters of Hudson Bay, the salmon or the whale returning to the Churchill estuary can still taste the trees of Thunder Lake.

Churchill, Manitoba, Canada
58° 46' 06" N

The towns of Canada's remote north use planes like very expensive buses. The First Air plane that delivers me to Churchill from Winnipeg must first fly to the barren settlement of Rankin Inlet on the north-western shore of Hudson Bay before turning and retracing its steps south again. The plane is divided in half – passengers sit in the back while the front half is reserved for cargo. The detour affords me a good aerial survey of the taiga–tundra ecotone that stretches from Churchill all the way to the Arctic Ocean. While in places like Greenland, the polar desert and the treeline are uncomfortably close, here the ecotone is 400 miles wide. The shrub line continues all the way to the far north of Nunavut's Arctic archipelago.

This is the place to argue over the definition of 'treeline'. According to one view, it is the northern limit of the growth of trees over five metres or more in height. But this seems a little cruel to the hardy and determined stunted spruce that cover this land in redoubts of krummholz. These trees are xerophytes: they are adapted to a xeretic landscape with strict limits on water, temperature gradients and reduced photon exposure. They are the fulcrum between the tundra and the treeline. In another version, the treeline is the northern boundary of the forest–tundra zone beyond the continuous forest where the balance between forest and patches of tundra changes to a majority of tundra over forest. Fixing that line at a point somewhere between Churchill and Rankin Inlet would be impossible without detailed studies of satellite imagery. Ask a hunter of caribou from the indigenous Sayisi-Dene nation, however, and they will know instinctively.

The Sayisi-Dene – known to the Europeans as Chipewyan, the name given them by their neighbours the Cree – call themselves the 'people under the sun' and refer to their traditional range as 'the land of little sticks', the flat plains of krummholz of the taiga–tundra interface. It is a large area. At its northern extremity, the Great Slave Lake in Northwest Territories, their territory overlaps with that of the Inuit. Before the Hudson Bay Company sucked them into its orbit of fur trading, they would follow the caribou from the edge of the boreal forest to their calving grounds on the tundra in the summer and back again in the winter. The dividing line between the habitats defined their calendar and their way of life.

A Sayisi-Dene folk tale tells how a long time ago humans and caribou lived together, but then some women tried to own the caribou, putting marks on their skin and ears with knives as the Sámi do. The caribou got angry and ran away. When they were eventually persuaded to come back they were much more wary of people. And indeed in many studies they are referred to as a 'bellwether' species, the most sensitive to disturbance. Caribou across Canada are now critically threatened. Few of the remaining herds are considered to be self-sustaining. Other species have started appearing in the ecotone.

'Moose are like cattle in the taiga these days,' says Dave Daley, an indigenous hunter from Churchill. So that's where the moose of

Pimachiowin Aki have all gone! And at least here in the Hudson Bay watershed the forest is intact enough so they can move through one continuous corridor. Such corridors are increasingly important, for as warming degrades habitats further south or makes things just too hot, animals' ability to move is already becoming a matter of life and death.

'The black bears are here, too, oh hell,' says Daley. The climate is right for them even if their normal habitat of trees has yet to catch up. Scientists estimate that the treeline is moving north here at the rate of about one metre a year, but over such a wide and varied terrain it is almost impossible to be certain. Big changes are taking place across the vast plain stippled by spruce and striated with water that unfolds below the aircraft at 10,000 feet. The skin of the earth is melting, microbial life waking after thousands, possibly millions, of frozen years. The soil is transpiring – perspiring one could say since more moisture is being released than absorbed – and animals and plants are taking note. It is a new world, and intelligent life – the smart genes – is sniffing it out, sending out suckers, seeds and scouts, ranging north, getting ready.

'They shouldn't be here.'

'What? The trees?'

'The balsam poplars. They shouldn't be here at all!' exclaims LeeAnn Fishback, a scientist at the Churchill Northern Studies Center.

We are driving in LeeAnn's truck down an esker – a gravel ridge created by the retreating Laurentide ice sheet. Meltwater running under the ice carved the moraine into ribbons that define the landscape all around the town of Churchill. It is one of the youngest landscapes on the planet, still in the process of being formed. The last ice age has a long tail. The volume and weight of the ice caused a depression in the crust of the earth about 270 metres deep. As the ice sheet melted, the water didn't escape to the sea right away, but sat atop the Hudson Bay lowlands, forming the inland sea of Lake Agassiz. Eventually, around 8000 years ago, the plug of ice holding the meltwater back dissolved, allowing most of the 841,000 square kilometres of Lake Agassiz to rush out into the ocean through the Hudson Strait in two successive pulses. Scientists estimate that this 150,000 cubic kilometres of water caused global sea levels to rise by one metre and inspired legends of floods in low-lying areas and the creation story of the Ojibwe nations of a land rising out of the water.[10]

Relieved of all that weight, the land is still rising at a rate of around three metres per century in a process called isostatic rebound. The effect is most pronounced along the shoreline. In the distance, grey rollers pound the gravel all along the flat rocky shore. The beach is getting wider by several metres every year as the ground rises and the water retreats. The wind carries stripes of rain across the bleak tundra scarred with the tracks of gravel-mining trucks and tundra buggies taking tourists on polar bear safaris. It looks lunar, and if the permafrost weren't melting the tracks on this delicate soil would last almost as long as the tracks on the moon.

The road across the esker takes us in a dead straight line inland, in the direction of the subglacial rivers, which trace the route of the retreating ice sheet in reverse. Away from the sea, the vegetation picks up. Willows, mostly, crowd the exposed gravel of the moraine. On the lower bogs on either side, spruce and tamarack (North American larch) huddle together on raised mounds called palsas. These rounded mounds of peat are formed by frost heaves as the peat expands and contracts, thawing and freezing. They contain a core of ice which provides moisture, but the dry peat around them is better drained and makes for a less waterlogged environment for trees, hence the appearance of tree islands throughout the fens. These are the healthy trees. In the bogs tamarack and spruce that used to enjoy the brief moisture of the summer as respite from the dry of the permafrost winter, are now waterlogged and dying from too much melt.

It's raining and cold. Twenty degrees colder than Poplar River only a few hundred kilometres to the south. This is the diversity of temperature range that Diana spoke about that makes the boreal so adaptable. It has experienced extremes before.

From our ridgeline vantage point, we survey the myriad ponds that speckle the tundra peatland – black discs of water trimmed with froth like shadowy hoods lined with fox fur. These are LeeAnn's speciality. The story of the peatlands can be read in the cycles of the ponds. As in Scotland, peat has been accumulating atop the bedrock here at a rate of a millimetre per year for 4000 years.

The forest has come and gone once or twice across this land, and the peat is where all that carbon is stored: 1.1 trillion tonnes of it – more carbon than humans have released by burning fossil fuels so far.[11] The flat, drizzly plain spotted with spruce and tamarack that continues uninterrupted in an endless sweep in all directions was once continuous forest. Soon after the retreat of Lake Agassiz, trees pressed into the breach up to 320 kilometres north of here, all the way to Rankin Inlet. Then it got colder 5500 years ago, and they retreated to their current latitude. Now they are mustering for an assault on their former territory, and trees that have been dead for 5000 years are beginning to rot again. That is what warming peat is: arrested decomposition, sequestered historical emissions, all being released at the same time. And that is why the warming of the boreal is so dangerous: it is not just the faltering ability of the forest to continue to sequester carbon that we should fear but the release of all the carbon previously sequestered by prehistoric forests as well. If Nadezhda in Russia is right and the tipping point has already passed when emissions from melting permafrost will now drive more warming regardless of what humans do, then we should be very worried indeed.

LeeAnn takes one hand off the wheel and points to a flat expanse of mud with a grassy lip rim. The ponds are drying out fast. The black mud absorbs more radiation, heating the ground further. Biological activity in the ponds is increasing – hence the froth – and when the ponds freeze, as they will do in a few weeks at the end of October, you will be able to see strings of pearls suspended in the ice – carbon dioxide, that enigmatic gas, made visible at last. The pearls are a product of the final decompositions of the

summer months, as breaking-down organic material uses up the
last of the oxygen. Then, as the water freezes all the way to the
bottom, larger bubbles appear lower down in the ice structure.
You can insert a needle into these bubbles and set the gas on fire.
It's methane, as in the Laptev Sea north of Siberia, where the anaerobic decomposition has continued in the sediment without oxygen.
LeeAnn's students enjoy that experiment.

LeeAnn stops the car. The wind slaps rain against the window.
It slides down in sheets.

'See that?'

A balsam poplar seedling, no more than five years old, nods in
the wind. Among the willow, spruce and larch it is certainly the
odd one out, its large leaves twirling precociously in the gusts. It
seems to be doing well: the stem is green and strong, the leaves
have a healthy colour. It is clearly loving the mineral-rich gravel of
the esker, and it is not alone. As we continue, more and more
juvenile balsam poplar line the road, competing with the willows
for the carbonate rocks eroded from the limestone of the Canadian
Shield. In one hundred years' time this track will be an avenue
lined with poplars dozens of metres high like a country road in the
south of France.

The esker comes to an end at a kame – a promontory where two
eskers meet and where the retreating ice left a widowed berg which
melted to form an elevated lake. The gravel here has subsided and
water stretches away on either side of the road. The residents call
the place Twin Lakes.

We get out of the truck into the gale. LeeAnn pushes two cart-
ridges into her shotgun and slings it over the shoulder of her bright
rain jacket. It is polar bear season in Churchill, the animals
marooned on land by the retreating ice and when females den in
the tundra to have their young. This low scrub with melting permafrost that is easy to dig is their kind of spot.

We push our way through the berry bushes almost overwhelming the stumps of spruce burned in a recent fire to reach the trees
that LeeAnn says should not be here. On a small incline above
the lake stand half a dozen impressive balsam poplars. They are

much bigger than I had imagined, reaching over twenty metres in height, their acres of leaves above me making a clattering noise in the wind. Up close the bark is very rough, cracked and furrowed into deep gullies, grey and black and slick with moss and black lichen. Four-inch-thick bark that doesn't burn and doesn't freeze. The inner bark has permeable membranes that permit the rapid movement of water out of the living tissue so that destructive ice crystals cannot form inside the cells. The trunk of the largest is broader than my chest. The leaf litter is dense, like a rotting carpet, and, pushing my hands into the dark wet layers, smells like one too. This is what feeds the underbrush and makes it so rich in juniper, willowherb, Labrador tea, blueberries, rose, wild gooseberries, raspberries and redcurrants. LeeAnn knew the place; it's where she comes to pick fruit to make jam.

It's a shock to see poplars this far north keeping very different company to their neighbours in the closed-canopy forest further south. Dispersal of species is still not well understood. Why are some trees here but not elsewhere? How did they get here? The white spruce, pollinated by wind, easily dispersed and easily germinated, is the natural leading edge of the treeline as it has moved over centuries and millennia. The balsam poplar finds sexual reproduction difficult. The small yellow flowers that emerge from the oily buds of female trees looking like the round heads of caterpillars breaking free of pupae every spring come before the leaves have even unfurled. They attract birds and insects of the boreal hungry for the first flush of nectar of the spring. When the flowers fall the entire catkin comes away. The fruit matures when the leaves are fully grown and splits in two to release tiny seeds, each bearing a tuft of long white silky hairs designed to catch the wind. The 'cotton' that gives the other poplars their nickname, cottonwoods, indigenous people used to use for spinning, for bandages and to line the cots of infants. The fluff was also added to buffalo berries and beaten to a froth to make 'Indian ice cream'.

That seed must fall on a moist seedbed, with easy access to mineral soil (recently burned sites are best) which must then remain moist for several weeks at the right temperature to allow for germination. But the seeds don't survive for long, and given the balsam poplar's penchant for rivers, flooding often puts paid to fragile seedlings in riparian areas. The fire interval this far north is over 400 years. A vagrant, fertilised poplar seed, even if it found its way here, would need to coincide with a fire event like the one a few years ago to stand a chance. Hence the tree's tendency to remain further south and LeeAnn's surprise at the presence of balsam poplars here at the treeline by the side of the Twin Lakes.

Steve Mamet, LeeAnn's colleague, has cored the local spruce, tamarack and birch and found the oldest to be around 400 years old. He has not cored the balsam poplars. I hope he will. Are they a recent freak accident? Are they refugees, the last outliers of their species from the time when the forest was further north? Or were they in fact planted by humans, and have been biding their time since, waiting for a warm spell like the current one to send out their suckers all along the gravel road? The eskers of Twin Lakes have yielded considerable archaeological evidence of an Inuit-related civilisation over 1000 years old that predated the Inuit 'Dorset' culture of the Middle Ages.

'It was a sacred tree, remember,' Diana said later. 'They could have brought it with them.' Just like the Celts and their sacred Scots pines. Why not? The poplar would have improved the habitat for fish, and the salt of poplar ash was used in preserving and cooking fish. In this part of the world just as elsewhere – in Scotland, Russia and Alaska – trees and people followed the retreat of the ice in tandem. Maybe they needed each other more than we know.

Churchill has always been a frontier type of place. The town was established as a fortified port for British trade with the First Nations and for the extraction of resources from the watershed of the Hudson Bay. King Charles II of England granted the Hudson Bay Company title to all the lands that drained into the bay, a territory known as Rupert's Land stretching into today's Minnesota and North Dakota. The fortunes of Churchill have always depended on the caprice of people far away.

The decision to run a railway line from Winnipeg to Churchill (along which ran a train with the unforgettable name, the Muskeg Express) defined the twentieth-century role of the town as a grain shipment depot bringing the harvest of the prairie to the sea just when the star of the Hudson Bay Company was failing. The decision by the US military to investigate the Northern Lights and the earth's magnetosphere led to a wave of scientists moving to the town in the 1950s and 60s, leaving a legacy of concrete launch towers in the middle of the tundra and the debris of failed launches scattered over Wapusk National Park. When I visit in 2019, the town is still smarting from another decision three years earlier to close the port by the corporation that owns it, Omnitrax.

In the town hall – an incongruous concrete edifice facing the Churchill estuary with an indoor basketball court, health centre, hockey rink, shuttered library and swimming pool open three days a week – the three staff whiling away their office hours tell me that the town has been in 'shell shock' ever since. The decision has recently been reversed, and the first ship in several years is expected soon, but the corporation's vacillation cost many jobs and at least one teaching position at the school. The people in the office talk about global warming with the same kind of resignation and sense of injustice: yet another misery inflicted from afar.

Further down the road, in a wooden cabin he built himself from spruce trunks cut at Twin Lakes, David Daley reclines in a wooden chair. The spruce he used were riddled with budworm – he wouldn't have cut live trees otherwise – and for the first year in his new cabin beetles poured out of the logs into the living area. He grew up hunting and trapping around Churchill and he's seen big changes in that time: the tamarack losing their bark in the swamps and not setting seed; the spruce decimated with budworm; birch getting into everything and balsam poplar sprouting all along the riverbank.

'The poplars have been that size ever since I was a little kid,' says Dave. He tells how the trees around Churchill were happy for many years but are now being claimed by the swamp as the permafrost melts. 'You can tell what is happening below ground by what's happening above.'

The poplars down at Twin Lakes, however, are content up there on the ridge. They like it, he says. Gravel doesn't freeze as hard as soil; it's warmer, less affected by the peat. Humans liked the ridges too. That was always the place for settlements. Dave has found seashells, whale bone and tent rings in middens on the esker not far from the poplars. It makes sense, he thinks, that his ancestors from the Cree nation, or possibly even further back, may have brought a seed or a seedling with them, leapfrogging hundreds of years of natural seed dispersal. By 'natural', scientists mean dispersal by animals other than humans.

We consider the balsam poplar invasion all along the river, and the possible regenerative effect on fish stocks the trees might have, but Dave shakes his head.

'Don't talk to me about fish! There's none!'

In this case the villain is clear. Another city corporation domiciled far away: a power company called Manitoba Hydro.

'I used to set my net and feed my dogs on sucker fish for days. Now there's nothing. Manitoba Hydro have killed the river. They let it get so low and then it froze to the bottom, killing everything. The dam holds back the sediment of the river, all the nutrients. No wonder the tour operators say the visibility is excellent for viewing whales! All the goodness has been strained out of the water.

'And there are no birds on the river either. I used to do a census on the river for Natural Resources Canada every year; now there's nothing to see. I complained to Manitoba Hydro but they told me, "We're not going to waste millions of dollars of water just so you can go fishing."'

Manitoba Hydro triggers massive flood releases every so often, but extremes of water do not a stable ecosystem make. Each time the layers of the river build back up again, and balsam poplar seedlings establish on the riverbank, they are washed away. There is a name now for the floods that happen when the dams release water – the volumes made larger by the linking of the Churchill and Nelson Rivers. Local people call them hydro tides. Poplar River flows into Lake Winnipeg, which used to reach the ocean through the Nelson River. But now the two rivers are joined, the minerals, nutrients and acids of Pimachiowin Aki and the other forests of the watershed that make up the unique chemical signature of the Poplar are discharged instead via turbines into the Churchill or the Nelson depending on how many people across Manitoba are turning on their air conditioning, boiling kettles or watching television that day. When Manitobans press a switch, they are probably not thinking about the water powering their comforts; the signature of the soil that connects the trees of the watershed with the fish of the river and the whales of the sea.

The upside of global warming is on full display in Churchill: POLAR EXPEDITIONS, SEA NORTH TOURS, WHITE BEAR EXPERIENCE, TUNDRA PUB, AURORA HOTEL and many other northern-themed tourist businesses are strung out along the customary wide single main street of the North American frontier, a street pockmarked with rusting containers, boarded-up enterprises and a single gasoline pump. The only traffic is an occasional SUV sporting knobbly tyres with deep treads or a huge 'argo' tundra buggy with wheels which seem to belong in an open-cast mine but were actually designed to take tourists across the wetlands.

Since the on-off closure of the port, the town has made an effort to rebrand itself as a tourist destination. The obvious strategy was to capitalise on the growing interest in witnessing the fragile landscape of the Arctic before it is gone for ever, and Churchill is blessed with the breeding habitats of two of the most iconic species of Arctic megafauna: the polar bear and the beluga whale.

Outside the Polar Hotel, I board a bus along with over a hundred other tourists from all over the world: California, Malaysia, China, Ireland and Australia. We take the road to the east of town, passing the municipal dump and grain silos as long and high as an oil tanker out of the water. The concrete towers loom over the town like an industrial overlord, reminding the inhabitants of their captive economy. Hundreds of metres up in the air an elevated conveyor belt projects out over the sea, ready to pour grain into the long-awaited ship. Built in the 1930s, the rusty steel and crumbling concrete silos dominate the estuary shore. Rough ground, fireweed and willows and ancient chain-link fences separate the facility from the marina where the bus deposits us at the end of a square of floating pontoons. We don lifejackets, listen to safety briefings and then embark in a fleet of black inflatable Zodiacs that purr over the milky water, the recently melted sea ice lending the estuary an opaque, shimmering, mineral blue.

The tide is in, the water dead calm. Mist rolls down the tundra on the far shore. The sky is the same shade of smoked grey as the sea. The Churchill River is half a mile wide, barely flowing, easing itself into the sultry sea with the minimum of effort. The Zodiacs circle on the surface like flies on a pond. I had thought that whale watching would be a pot-luck activity with no guarantee of a sighting, but my fellow passengers are on their feet almost as soon as we join the main stream, 'Over there! Over there! And there . . .' In every direction curving white backs breach the surface like slabs of cream. The river is churning with beluga whales, *Delphinpterus leucas* – the white dolphin without a wing.

Our guide pilots our inflatable in loops around the feeding whales. There is a pod for every boat and more to spare. They thread themselves around us, under the boat, rolling on their sides to get a good look at these strange spectators of their lunchtime. Of all the baleen whales, the beluga is the only one with a flexible neck. Lacking the fused vertebrae of its cousins, it can twist and turn, giving it even more of an anthropomorphic air. Its curving beak is fixed into a grin, and the echolocation equipment housed in a bulb called a melon on its forehead causes a furrow resembling eyebrows which in turn makes it look like a curious geek. They are small whales but still grow to over four metres – sizeable, intelligent beings, longer than the boat and outnumbering us by ten to one. For a while five or six of them nuzzle in behind the Zodiac, enjoying the wake. The guide isn't sure if they like the warmth of the exhaust or the oxygenating effect of the churning propeller, or if there's something compelling in the frequency of a ninety-horsepower Yamaha engine.

The climax of the tour is the lowering of a hydrophone – a waterproof microphone – behind the boat. A speaker on the floor of the boat erupts into pips and trills, whistles, long squeaks and rhythmic clicks. The belugas are having a heated conversation down below, and hearing their own voices feed back drives them into a further frenzy of chatter. They have 1200 different sonic signals, a far more sophisticated alphabet than *Homo sapiens*. Belugas have been shown to raise their voices to be heard, just like humans, but at a certain point they give up. Their stress levels correlate closely to the level of noise pollution. The massive expansion in Arctic shipping projected to accompany the end of sea ice may be the most deadly threat to the belugas, who 'see' with echolocation and who exist in an almost permanent web of sound. They are constantly talking to one another; to live is to chat. There are literally whole languages that humans have yet to understand.

More come, crowding the rear of the Zodiac, craning for a look at the floating humans who are mimicking their debate. A mother joins the edge of the pod. Riding on her back is a calf about the size of an adult human still with the tatters of its first orange skin. The belugas calve in the shallow waters of the estuary, which are comparatively warm and more oxygenated than the open ocean. Females gestate their young for twenty months and nurse them for up to three years, as they whiten with age and milk. Specimens born in captivity have never lived longer than thirty years, leading researchers to initially believe the lifespan of belugas was strangely short, although it is now thought they live to over a hundred. This calf was probably conceived here two years ago. Belugas, like salmon, always return to the same river.

It makes sense that the homing instinct might have evolved as a hormonal–chemical lifeline back to a safe place where reproduction can happen in relative peace and with an abundance of food. But such a finely attuned and intelligent plan now looks like a liability as climatic zones and ocean currents become unpredictable. The opportunism of the explorer or the colonist is a different skill set, a different mental geography than the genetic call of home. Will the belugas and others be able to learn in such a short space of time? Will there even be a climate analogue – the scientific term for a location that matches their current habitat and range – in a world three, four or five degrees warmer? As all the species of the boreal move further north, those at the highest latitudes like the beluga and the polar bear have nowhere else to go.

'So what's so special about this place?' I ask our guide. 'Why do they keep coming back here?'

'There's a few theories,' he says. 'Moulting their coats on the rocks in the river; it's shallower; killer whales can't get into the estuary, and there's lots of food. But basically, we don't really know.'

I think I might have a clue. The pod following us breaks up when the hydrophone is removed, and the belugas go back to slicing the water in gangs, hunting capelin, *Mallatus villosus*. Capelin are so abundant at this time of year that masses of them are washed ashore on the glacial beaches. Residents fill buckets and others complain about the stench. Our guide doesn't say this, but the capelin are here because of the plankton, and the plankton and phytoplankton upon which they feed are here because of the sweet spot where the gift of the melting sea ice meets the gift of the trees flowing downstream.

The beluga only venture south into the estuaries to calve during the brief summer melt; they live the rest of the time at the edge of the pack ice, breathing in stress fractures formed by currents called polynyas and leads. In extremis they can use their hardened bulbous heads to break through the ice to take a breath. The 'dolphin without a wing' evolved without a dorsal fin to swim under the sea ice, feeding at the lower trophic levels of the food chain on minnows, smolts, shrimp, plankton and crustaceans, all of which depend on the nutrients that come down the river or slowly fall out of the sea ice like rain, 'seeding' the ocean.

The critical role of sea ice as a platform for marine life at the bottom of the food chain has been known since the 1960s, and yet its rapid melting is still viewed by many as just the unfortunate loss of a picturesque environment or, worse, an opportunity for new shipping routes. It is in fact a catastrophic weakening of the ocean food chain – the marine equivalent of removing huge amounts of topsoil on land.

During the winter the heavy (liquid) brine drains out of the sea ice, falling to the ocean floor to cause a circulation of ocean water, bringing nutrients to the surface. Within the freshwater crystals of sea ice are channels vacated by the salt in which diatoms, micro-organisms suspended in the ice, can begin to thrive. As soon as the sun hits the pack ice after the polar winter, the ice crystals attenuate the light and plankton take the iron chillated in the humic acid identified by Professor Matsunaga and Diana and begin to divide.

All through the spring the plankton continue to divide in the ice, peaking at break-up when, finally released from their crystal cocoon, the injection of freshwater and its crucial load of iron into the sea allows the plankton to divide like crazy, providing a banquet for the capelin and other smolts, the minnows and hatchlings. It should be no surprise that this activity is busiest at the mouths of the major rivers. The essential fatty acids in fish – oleic acids, used for producing eggs and milk – come from phytoplankton, which recruit it from minerals dissolved in freshwater whose origin is the leachate from decomposing leaves. One month after the melt and the plankton bloom, along come the belugas for their feast.

The foundations of the feast are fragile: there must be enough trees in the watershed, not too much agricultural pollution, and the sea must not get too warm. The invisible forest of the ocean – the underwater and intertidal algae – needs a critical temperature gradient in the surrounding sea (cold water compared to the heat generated in the plant) in order to reproduce.

The vanishing sea ice will change things for the belugas. It might also change the circulation patterns of the ocean. History shows currents have altered before, and it can happen as quick as a light switch. Citing geologist Alan Morgan of the University of Waterloo, Ontario, Diana says it can happen in two weeks. All of a sudden, one day the Gulf Stream, for example, could go into reverse. There is chaos in the oceans right now, so no one can predict how temperatures and currents and ranges will shift. Computers cannot do accurate modelling because the variables are too great. And the feedback between the warming of vegetation on land and the chemistry of freshwater drainage on primary production in the ocean is a great unknown.

The belugas have long been nicknamed the canaries of the ocean because of their melodic chirping songs. But, like the trees to which they are connected, they are now also canaries in another sense. When the critical productive relationship between the forests of the land and the forest in the sea breaks down, the belugas will be the first to feel it. For now the scientists monitoring belugas are content. There are 57,000 in Hudson Bay, the number stable, with the largest concentration in the Nelson River, the main outflow of Pimachiowin Aki.

But Dave Daley is not so sure. In his log cabin made of worm-riddled spruce he presses his lips together and shakes his head. He knows what happened to the belugas at Kotzebue: they disappeared from one year to the next. 'They say the belugas are holding up. But . . . there isn't the capelin there used to be. And these whales live for a long time. I don't know. I'm waiting.'

These graceful, playful, joyful animals arcing around the boat know far more than us about what's going on down there. Using infrasound they can communicate over thousands of miles. Like Albert talking to people in British Columbia in the shaking tent, all the belugas in the sea might be talking to each other all the time. One day they might hold a virtual conference call and the next be gone.

6. Last Tango with Ice

Greenland mountain ash, *Sorbus groenlandica*

Narsarsuaq, Greenland
61° 09' 41" N

Kenneth Høegh planted his first tree at thirteen. On Saturdays and after school he had a part-time job working in the children's section of the local library in Narsaq, in the far south of Greenland. By his own account, he was a bookworm. One day he came across an article in a science magazine for children about tree-planting experiments in a sheltered fjord further north.

'Growing up in Greenland, trees were kind of exotic, strange,' he says.

There were none in the small town of Narsaq, an otherwise picturesque assemblage of pastel homes on a green hillside arranged around a natural harbour that froze in winter. He had encountered very few trees in his short life. He thought he might like to try his own experiment and so he asked his parents to buy him a seedling, which came, airfreighted, from Iceland, and which he planted in his parents' garden. It was a Siberian larch, *Larix sibirica*. His parents moved, and the garden now belongs to neighbours, but the larch is still there. Today it is five metres tall and Kenneth is fifty-three.

Kenneth went on to study agronomy at university and became an agricultural adviser to farmers in southern Greenland while continuing to indulge his passion for planting trees. The difficulty for would-be foresters towards the end of the twentieth century in Greenland was that there were almost no trees. However, there were two forestry experts. Since the 1970s, ecologists Poul Bjerge and Dr Søren Ødum had been running experiments in the valley of

Narsarsuaq, to see what kinds of trees might grow in Greenland. It was their work that Kenneth had read about, and he went to work with them.

During the 1980s and 1990s Poul, Søren and Kenneth travelled to many locations along the Arctic treeline to collect hardy specimens of boreal species to bring back to Greenland. They went to Alaska, Yukon, British Columbia, Hudson Bay, Quebec, Norway, and across Siberia from the Ural Mountains to the Altai, to Kamchatka and Sakhalin, in the process assembling one of the most comprehensive arboretums of boreal treeline species in the world. With 110 species so far, it rivals that of the Sukachev Institute in Krasnoyarsk. The scientists' aim was to establish a reference for Greenland. As global temperatures rise, it is a resource with significance far beyond the confines of this narrow clutch of fjords: a bellwether for the northern forest and a future refugium for species driven to the brink elsewhere.

Kenneth loved the science of comparative research and long-term study of individual species, but he also just loved planting trees: tens of thousands of Siberian larch, Engelman spruce, Norwegian spruce, white pine, lodgepole pine, Douglas fir, balsam poplar and many others. What is now the national arboretum, the Arboretum Groenlandicum, *Kalaallit Nunaata Orpiuteqarfia* in Greenlandic, is a young forest covering one half of the valley of Narsarsuaq with over a quarter-million trees.

As the plane makes its steep banking turn into the mouth of the fjord, the blinding white and blue are divided by a thin wedge of forest glowing green and yellow with the odd burst of blood orange. It is August 2019. The late summer foliage is a flash of life in an otherwise barren landscape of rocks and grass caught between the ice cap and the sea.

The same topographical conditions that make Narsarsuaq a good place for planting trees made it an ideal place, almost the only place in southern Greenland, flat enough for a runway. If it was up to the Inuit, Narsarsuaq would not be a settlement. A wide, open valley without a natural harbour far from the summer sea ice has little attraction for a people that rely on the sea. But the United States Air Force, which built the runway, wasn't looking at it from a subsistence point of view. In 1941 the Germans had just invaded Denmark, and the Americans were looking for a staging post, a place to refuel the fleets of B-17 aircraft on their way from Georgia via Canada and Denmark's Greenland colony to Scotland to fight in Europe. All the supplies and construction materials were shipped or flown in, including 5000 troops who went on to build the Cold War base known as Bluie West One.

The Danish government saw no need to mess with the infrastructure they inherited from the Americans in 1958, and nearly seventy years later Narsarsuaq is still the only international airport outside the capital, Nuuk, the gateway to what the tourist board calls 'Sunny Southern Greenland'. Visitors are treated to an aerial moment akin to the closing scene of Stanley's Kubrick's movie *Dr Strangelove*: planes arriving from Europe skim over the blinding ice cap between serrated black peaks, bank at ninety degrees above a turquoise fjord dotted with icebergs and then descend steeply to what has been called the world's most dangerous runway, with one end running down to the water and the other abutting a cliff over a foaming glacial river.

My phone pings as soon as we have navigated the terrifying landing and climbed down the aeroplane steps onto the tarmac surrounded by majestic mountains on all sides: 'You'll find me in the hotel restaurant.'

On the ground in the rudimentary terminal with a broken baggage carousel, dozens of tourists with large brightly coloured backpacks and anoraks jostle each other. They are keen to hike the wilderness and see the ice before it is gone. This is the last summer before the pandemic, and business is booming: the year before there were 92,677 air passengers.

Some visitors are waiting for helicopters to take them on to the other settlements of southern Greenland, others are exiting, like me, onto the American tarmac road that leads into the town, if it can be called that: a handful of houses, two cafes and four barrack blocks repurposed as government-supported housing and painted

Greenland's signature pale blue and lurid green. The road slopes gently downhill towards the settlement and the sparkling blue of the fjord ahead. The buildings huddle under a knobbly ridge dotted with saplings that ends in a promontory dominated by a clutch of white cylinders squatting like sentries above a small harbour and a muddy beach lapped by dark blue waves. The cylinders are marked JET OIL 1, JET OIL 2, DIESEL and KEROSENE – another American bequest.

On the steps of one of the housing blocks a man is cleaning a rifle barrel. He looks at me and then carries on polishing. An elderly Greenlandic woman in slippers is pinning washing to a line in the sunshine. Two children race around a forlorn-looking concrete playground. On a gravel terrace above them a white windowless plastic box houses the supermarket. Outside, two women in parkas are selling kitchenware from China and hot dogs grilled on a gas barbecue.

Next to the playground is what I have been looking for. The Narsarsuaq Hotel is another repurposed military building. Beyond the open-plan reception area, the ghosts of US servicemen linger. The stainless-steel canteen without windows could be in a military base anywhere in the world.

At a table, sitting amid the debris of their lunch, I find Kenneth, with short sandy hair, and Peter, a Danish scientist with piercing blue eyes and points of orange light in an impressive beard. As soon as I arrive, we are off, out the back door, into blinding sunlight, past the rubbish bins of the hotel and onto a patch of waste ground once used by the Americans as a gravel pit. The sun is hot and especially brilliant. About a dozen people are scattered over a patch of thin grass busy with what look like rocket launchers aimed at the ground. To one side a pickup truck is half unloaded with stacks of cardboard boxes. A tall man in a pork pie hat with exotic feathers sticking out of it, ginger goatee beard and blue sunglasses grabs a box off the back of the truck and walks into the centre of the group.

'Yo! Poplars everyone! Grab some poplars!' he shouts in an American accent.

People crowd around, take handfuls of seedlings and then wander off to feed them into the open chambers of the bazookas. I
drop my rucksack, take a handful and find a partner. When I turn
round, Kenneth is gone. Before long I am on my knees planting
balsam poplar seedlings with a Spanish man called Miguel. The
sun warms my neck; people are sweating, taking off their coats
and jumpers. The ground is hard. The stony cliff at the edge of the
valley shimmers slightly above us. A background roar of river is
just perceptible, like a motorway over the hill. To the east, trees
cluster on the prow of the ridge, the beginning of the arboretum.
To the west, housing blocks begin.

'Don't plant too far that way!' shouts the man in sunglasses,
who appears to be in charge. Apparently, the inhabitants of the
blocks don't want too many trees spoiling their view. The man
with the rifle is still sitting on the step, the weapon across his knees.
An occasional passer-by glances our way and then continues on.
Foreigners planting trees in a frenzy on waste ground at the edge
of town is apparently an everyday sight. But the local inhabitants
are not involved. Trees, once essential to survival here in the form
of scrub and driftwood, are no longer a determinant of human life.
Now, the crucial factors are the flights from Denmark, the ships
that bring food, alcohol and fuel from Iceland and Canada, and
the presence of fish and game to hunt. Polishing a rifle is more useful than planting a tree.

Miguel and I plant a whole box, fifty trees, while the American
marches around filming the enterprise on his iPhone. Miguel
speaks little English but I gather he works with tourists leading
groups onto the ice cap. Our fellow planters include a man who
looks Japanese, a woman and child who seem Middle Eastern, several white men and one or two Inuit. I have failed to grasp from
Miguel what is going on. The man with the iPhone is wearing a
T-shirt printed with GREENLAND TREES. He disappears behind
the truck and just as suddenly comes back with another cardboard
box.

'OK, drink! Sodas! Hey everyone, it's freakin' hot!'

The American hands me a Coke and introduces himself. He is

Professor Jason Box, a climatologist with the Geological Survey of
Denmark. Greenland Trees was his idea, he tells me, although he
doesn't seem to be planting too many himself. Instead he talks,
waving his hands around and gesticulating at the mountains, the
fjords, the people planting. He is a constant, whirling ball of energy.

'Some people say, just leave it, let nature do her thing, you're
bringing in invasive species. The ecologists don't like us. But I say,
fuck it! We're helping nature along. Seedling recruitment here
takes a long time, there's no natural source of seed . . . Peter and
Kenneth tell us where to plant.'

Jason starts pointing: Dirk and Maurice from the Netherlands,
Masahito from Japan, Chris from the United States, Fazia from
Iran, all experienced and well-respected climatologists engaged in
research for the UN Intergovernmental Panel on Climate Change
(IPCC), currently getting dusty knees pushing balsam poplar seed-
lings into the thin soil.

The project started off as a way for guilty glaciologists to offset the emissions of their research projects; airlifting research
equipment and tons of gear up onto the ice cap each summer
requires a lot of aeroplanes, boats and helicopters. But, quite rapidly, other scientists came on board. They were so traumatised by
the extreme melting they were seeing on the ice that they wanted
to do something tangible and urgent about removing carbon dioxide from the atmosphere. They looked around for where to plant
trees, then they remembered they had been flying over the arboretum for decades.

In the evening, after planting 1300 trees, all the volunteers are
invited to a barbecue of reindeer and Greenland lamb. Further up
the valley, beyond the old base, at the entrance to a smaller glen
that the Americans called Hospital Valley, is a small wooden cabin
now belonging to Kenneth with a flagpole long relieved of its Stars
and Stripes. Kenneth planted the conifers surrounding the cabin
twenty years ago; they are just beginning to impinge on the faint
blue of the dusk. Once the sun has travelled over the mountain the
temperature suddenly drops close to freezing. While in daytime it

can reach up to twenty-five degrees Celsius, the proximity of the ice means that Greenland nights, even in summer, are always sharp.

The scientists warm their hands around the bonfire and drink beer. They are dressed in the uniform of their trade – technical expedition clothing emblazoned with logos signifying glamour: EXTREME ICE SURVEY, DUTCH ARCTIC EXPEDITION. The Dutch are disproportionately represented. A quarter of the Netherlands is below sea level, and melting glacial ice is the number one threat to the country.

Peter is the only one not in branded clothing. A Danish climatologist, he studies the historical patterns of vegetation. His outfit is that of a forester: shirt, trousers, fleece and coat in contrasting shades of green. He sits quietly as the younger scientists on the other side of the fire worry about the weather. At dawn they are due to rendezvous with a boat and a helicopter at the edge of the ice cap and travel with a BBC film crew to check on the steel poles they planted in the ice last summer. The depth of the poles will tell them how much the ice cap has melted during the last year. It is the final weekend in August, the end of the melt season, usually. But heavy rain is forecast, and they go back and forth mapping out options with military intensity.

'Ground truth' is what Dirk, a Dutch glaciologist, calls the work they are doing. US Air Force planes have flown surveys across the ice cap with radar to try to gauge the mass of ice and the rate of melt, but they can only measure in a straight line. Satellites equipped with lasers have been sent into space to do the same thing, a NASA project called GRACE: Gravity Recovery and Climate Experiment. NASA measured the difference between the earth's crust and the surface of the ice cap and estimated the mass of ice. Every year 300 cubic kilometres is being lost, and it is accelerating.

None of the methods is perfect, and their estimates must be supplemented by actual measurements from the surface of the ice. Some of these people have been coming here for nearly two decades; for others it is their first trip. The number of projects in Greenland – and the number of scientists – is expanding rapidly.

More important than tree rings, the climatic archive contained in prehistoric ice is critical to our understanding of how earth's climate has changed in the past. And the rate of melt of the ice sheets is pivotal to what will happen next. There are two main ice sheets on the planet, Antarctica and Greenland, and Greenland is much easier and cheaper to get to. It is also melting much faster than Antarctica. Greenland is at the front line of glaciology and sea level rise, and the people around the fire are responsible for much of what we know about how the disappearing ice will change our planet.

Dirk built the network of measuring stations for the Antarctic ice sheet; Jason built the one for Greenland. The data feeds into the complex climate models that other scientists such as Masahito construct. We talk for a while about models. This is Masahito's second trip to the ice cap this year already. It is a long way from Japan, where his lab is constantly refining one of the most sophisticated models in the world, one relied on by the IPCC. He recognises that the model struggles with climate feedback loops, and from the way he talks, rapidly, with wide eyes and hunched shoulders, it is clear he is obsessed about plugging the gaps. When I mention that some ecologists believe earth systems are so complex that they are impossible to model, the conversation stops dead.

Once upon a time models were tools for scientists, but it seems these days that humans have become tools of their models. Research grants are poured into projects designed to gain more data to refine the models built with supercomputers. But models are dangerous – they can be adjusted depending on the story one wants to tell. The fifth IPCC report of 2013 famously relied on models for the previous ten years instead of actual available observations that painted a far more pessimistic picture of Arctic sea ice melt.[1] You can't argue with real data.

Real climate data is the most valuable currency on earth at the present moment, and these scientists are modern-day treasure hunters, undertaking dangerous and glamorous work with helicopters and ropes, sleds, state-of-the-art technology, braving crevasses and blizzards to bring back priceless ice cores into which they stare in order to divine our future. But measuring, witnessing and dispassionately explaining the objective probability of ice sheet collapse and what that means for human life on earth creates an emotional contradiction. Scientific research projects always conclude by identifying areas meriting further research to contribute to the total sum of knowledge. Western science is a product of its own telos, an ideological idea of progress that now seems like a redundant cult of futurism. It assumes that there is always more time.

While the younger scientists like Maurice seem driven by an urgent search for decisive facts, the older ones are more cynical. Year after year, documenting ever more dangerous levels of melting while seeing no action breeds despair. It was why Dirk and his wife Fazia, also a leading glaciologist (who famously documented the acceleration of the Kangerlussaq glacier over a decade previously), decided to quit academia and devote themselves to Greenland Trees instead.

These days they look at Greenland differently. Before, Narsarsuaq was the place to wait for a helicopter up onto the ice or get drunk in the bar of the Narsarsuaq Hotel swapping professional gossip with other scientists passing through without ever meeting any local people. Instead, tomorrow, when Jason and his team go up, Dirk and Fazia will take a boat south with Peter to talk to schoolchildren about the planting project. They used to see only the white part of the island. Now they look at its potential to turn green.

The scientists have gone back to the hotel. The fire is a handful of embers. In the half-light a luminous glow persists behind the dark mountains outlined in relief, and the river's faint roar like a distant plane animates the night. The air is cold and crisp. Greenland is an anomaly in time and place. At sixty-one degrees latitude, Narsarsuaq is a long way south, a comparable latitude to the Shetland Islands, the middle of Norway and Sweden or Anchorage, Alaska, all subarctic boreal locations. But the East Greenland Current, which

brings sea ice down the coast from the Arctic Ocean, and the cooling influence of the ice cap dictate its peculiar microclimate.

'It is a very curious thing,' explains Peter. He is referring to the
fact that Greenland's inland fjords are an ideal climate for boreal
tree species, yet there are hardly any trees.

The average yearly temperature is well above freezing, and lately
the summer temperature has been pushing above Humboldt's July
ten-degree isotherm, the traditional definition of the limit of northerly tree growth. The new normal for July, says Peter, is above
eleven degrees.

But the reason there are no trees in Greenland is not because it
has been too cold. It is because there has been no seed. Vegetative
changes take thousands of years. This is what ecologists call disequilibrium dynamics, an ecosystem or a biome that is in the
process of working out its equilibrium. Rather like the changes
under way with our current global heating, there is a time lag as
species and ecosystems catch up with changed temperatures or
currents – in some cases thousands of years. Greenland is still playing catch-up from the last ice age.

The previous glacial maximum covered the whole of the island;
now, the ice has retreated to approximately 80 per cent. Without
the influence of humans, ecologists estimate that the 'migration
lags' of tree establishment could be thousands of years. And this is
further complicated for Greenland by extreme variations in eleva-
tion and topography among the mountainous fjords. The species
that have arrived so far are those that can be dispersed by air.

Very light seeds, such as those from the catkins of birch and
alder in Labrador, have been carried on the wind and settled in the
fjords. These, and the seeds contained in the berries of rowan and
juniper propagated by birds, have given rise to the only four current native trees of Greenland. The common juniper is found
throughout the boreal. The birch and alder are North American
varieties easily traced to the neighbouring continent across the
Davis Strait. But Greenland mountain ash, a kind of rowan,
proper name *Sorbus groenlandica*, is a strange subspecies believed
to have colonised North America from Greenland during the last

interglacial. It has now found its way back again from a refugium across the sea.

It is low-growing and much smaller than its European counterpart, *Sorbus aucuparia*, or its two North American cousins, *Sorbus americana* and *Sorbus decora*, with tiny versions of the rowan's fronded leaves and the same silvery bark and distinct scarlet berries with a five-pronged pentagram at the base of the fruit beloved by birds and previous generations of humans alike. The story of the Greenland mountain ash is a good example of speciation, the mysterious process of evolution governed by the rhythm of planetary trajectories, the pulses of ice ages and the imperceptible unfolding of geological time. And it is a fable for the species of the boreal portending the challenges to come.

Some time between previous ice ages, a species of rowan tree ended up in Greenland and seeded a local population. *Sorbus* are found all around the boreal region, from Scandinavia to Siberia, and everywhere have shown an ability to adapt by hybridising vigorously. The capacity to hybridise is a survival strategy, a useful skill for the Anthropocene, and the rowan is a survivor par excellence.

The bark is smooth and silver, reflecting sunlight that might thaw its super-cooled sap. Its leaf buds are purple, dark boxes to attract solar rays and send out the first flowers in spring. The rowan is a hermaphrodite. Each creamy white five-petalled flower contains both male and female parts. A member of the Roseacae family, its clusters of tiny flowers are miniature replicas of roses, holding both dark pollen and one of the first nectars of the year, welcoming insects great and small. The rowan cannot afford to be fussy about its visitors, since there are so few. The sorbic acid it offers is critical for wasps and moths and other insect life in the spring, and songbirds seek the sugary sorbitol in its distinctive red berries in the autumn. The rowan is a fixture of the calendar, like a boreal clock: the white blossom is a harbinger of spring, the reddening of berries the flag of autumn and their disappearance a sign that winter has arrived.

Disconnected from other populations, perhaps by the sea or

more likely by ice, the Greenland population kept its own time, evolving to suit its new habitat, producing smaller leaves and becoming less ambitious in height. The same features of the Greenland topography that limit seed dispersal make it ideal as a possible micro-refugium.

In the past, mountains and islands have served this function. Mountainous places with varied topography like Greenland have wide variation in habitats and can also see inversions of atmosphere – where warm air is trapped in the valleys and kept in place by cold winds above. Geological records hint at sanctuaries like this in valleys spared by the ice where species have huddled together in odd combinations with unpredictable results.[2] Geologists call ice ages species pumps, because of this cyclical forcing of evolution.

When the ice retreated and birds found their way across the sea once more, the Greenland mountain ash was transplanted back to find the original population had evolved beyond recognition. It was now its own subspecies. In other parts of the world species ranges have diverged but never overlapped again. In the Appalachian Mountains of the United States relict boreal species are related to others now found in the far north of Canada. In Tanzania now separate cloud forests on the tops of tropical mountains such as Udzungwa or Usambara contain endemic species marooned when the forests that once linked the habitats retreated and became savannah. The Himalayas are another example.

After the coming climate chaos, once the earth has found a new equilibrium what species will remain? The answer to that question will be determined by refugia: whether or not they can offer enough of a haven for long enough, and whether or not they can be reached at all.

The harsh mantra of a warming world is 'Adapt, move or die.' But some species can move more easily than others. Climate change velocity is the pace at which a species will need to move to track its climatic niche. For conservationists, hard choices will need to be made, like Noah and the ark, although the ark is nowhere near big enough. This is the emerging field of strategic ecology. Planting based not on the current climate but on guesses about the future.

On the equator, the warming planet will eventually chase the endemic species of the cloud forests up beyond the summits into the thin air of extinction. In the northern hemisphere, species are already trending north. For mammals like the migrating moose, caribou and bears of northern latitudes, or the birds of the Taimyr peninsula, moving is no problem, provided there are contiguous corridors of habitat and food sources, at least until the Arctic Ocean. But what if you are rooted to one spot? What if you are a tree?

Scots pines for example usually establish new seedlings less than 200 metres from the parent tree with the odd rogue seed travelling further afield. Pine forests move in packs. Some species, like *Cerbera odollam*, sometimes called the suicide tree, only have one offspring, *in situ*, fed by the rotting carcass of the mother. Such specialised species already have an unsustainable climate change velocity, which means they will go extinct without human assistance. A temperate or tropical tree currently facing heat stress might have a climate velocity of several kilometres a year.[3] At that pace, if the tree could walk, you would be able to see it move.

There are *in situ* refugia, which species might reach on their own, and then there are *ex situ* refugia – places well suited to ride out the storm but to which species might need to be transplanted – what is called in the literature assisted migration. Such technical climate phrases are set to enter the popular lexicon – dry, clinical language masking an implicit surrender: we failed to stop it.

Kenneth did not start planting trees with climate breakdown in mind. He and his colleagues began with a very basic insight: in North America *Sorbus groenlandica* is not usually a treeline species at all, therefore there must be scope for others on Greenland. But the question they were asking is essentially the same as the one currently obsessing conservationists and strategic ecologists in the scientific community: what is Greenland's climate analogue? What else could grow here, if it had to?

They were not the first to ask these questions. In 1892, in the spirit of colonial experimentation to see what resources could be gained from Denmark's northern possession, Danish botanist L. K. Rosevinge established a test plantation of six Scots pines further up the fjord from Narsarsuaq.

The finger of sea is called Qanasiassat in the Inuit language, and it is about as far as you can get from the open sea in Greenland's coastal archipelago. A boatman called Jiaggi takes me up there in a speedboat early in the morning. The first snow of autumn has fallen in the night, anointing the mountains that frown upon the farms of the few hardy people attempting to make a living from sheep on the steep slopes. Rags of cloud cling to the summits, and the surface of the fjord wears a mist resembling fur that is the colour of dirty milk. The river rushing down the valley of Narsarsuaq is loaded with sediment from the melting glacier. Pockets of melt accumulate under the glacier, Jiaggi explains, and periodically they overflow, causing the river to flush with cloudy particles.

The harbour left by the Americans is still intact, its huge trunks of Oregon pine tarred with pitch lashed together – trees the like of which Greenland might see one distant day. The heavy boat cuts a white gash in the perfect cloth of the fjord. The sun is still behind the hills, but even at this time the sky carries a backlit glow from the ice just beyond the horizon. Jiaggi is French. He came here on a home-made boat with a group of other hippies in 1976 and was the only one who forgot to go home. Instead he started a tourist business called Blue Ice Explorer that is just, as he retires, booming, with people attracted by the fatal pull of that disappearing blue ice. He shrugs. Next year, he and his Danish girlfriend are buying a camper van and planning to return to his native Alps.

At the head of the fjord the sea meets an amphitheatre of spectacular rock carved in an almost perfect circle. Onshore, the plantation looks odd, like a square of velvet pinned to the mountainside. Inside a broken barbed-wire fence is a hectare of more recent plantings: larch, white spruce, jack pine, Scots pine. Rosevinge's original pines stand, windblown and krummholz'd at the edge of the green square, wizened and grey. Only two seem to be still alive, bark stripped by the wind and remaining needles brown with flecks of green only at the tips. The grass all around is thick and spongy, clogged with moss. For 120 years people have been coming to pay homage to these refugees from northern Norway, now finally, it seems, nearing the end of their lives. Bugs have clearly ravaged one of them. Sheep have nibbled at some of the lower branches.

In the Narsarsuaq museum there are photographs of happier days. Rosevinge's mistake, explained the curator, Ole, was to choose specimens from northern Norway that he believed to be from a similar climate. In fact, the genes of the northern pines were coded to expect much less daylight and shorter seasons, so they were not adapted to take advantage of southern Greenland's longer growing season, shutting down in early autumn when other trees were still soaking up the sun.

Next to the sickly pines is a very impressive larch tree with broken sweeping branches that looks to be over one hundred years itself. A metal tag carries the number 3799. Inside the square of man-made forest there is a quiet hush. A carpet of needles is rapidly decomposing into a rich humus that gives off a pungent scent. The sound of every forest is distinct – the complex of leaves and needles filtering the air into a unique susurration of rustling leaves. This forest sounds like quietly undulating static.

At the far end of the plantation is a fully equipped cabin with firewood, tools, cooking utensils and blankets. A lonely place to be marooned but not a bad one. The view down the fjord of icebergs and white frozen peaks beneath purple rain clouds and the occasional cruise liner would be a fine backdrop to the end of the world.

Once I have walked all around and through the hectare, there is little else to see and it's time to wave to Jiaggi moored out in the fjord and re-board the boat. It is only when looking back at the square of dark evergreen that I realise what is strange about the plantation. There are almost no seedlings, even within the fence. Boreal species need fire or disturbance to germinate, they need mineral soil, not soil covered in wet moss and sedge. Outside the enclosure, the mountainside is dominated by low-growing birch, *Betula glandulosa*, the native hybrid. I spot only one spruce seedling on a scar of bare earth above, where a landslip has exposed the ground.

The timescales of natural afforestation at high latitudes are suddenly clear, and it's frightening. It takes centuries, if not millennia, for a forest to take hold, even where climatic conditions are right. The soil, rainfall and disturbance rate must all be aligned. Left to its own devices, nature has decided that birch is the appropriate species here. The birch might prepare the soil – or burn – clearing the way for conifers, but not yet. Although the conifers in the plantation put in later than Rosevinge's pines appear healthy and thriving for now, the plantation's lack of regenerative capacity is actually an argument against such projects: without human intervention or disturbance, these trees will not reproduce on their own. They will remain a relic on a hillside until, in the end, something else takes over.

The boat swings around the head of the fjord away from the blank grey slope of scree and towards a majestic southern vista of thunderheads, mountains, snow, ice and sea. The natural habitat of mosses, sedge and birch stops at a certain height and gradient. Slabs of perfect homogenous green appear to have been scooped out of the hillside. In these miniature fields, islands in the brown and grey, tractors crawl at unfeasible angles collecting huge wrapped bales with which they construct pyramids of black plastic. The sheep farmers of the fjord are stockpiling hay for the winter, just as their Norse forebears did on the same fields 1000 years ago. Part of the reason there is so little native birch forest in southern Greenland these days, part of the argument for test plantings, is that the Vikings cut it all down.

On our way back, Jiaggi wants to deliver the post. Opposite Narsarsuaq, across the fjord, is an older, more populous village called Qassiarsuk. When the Americans came in World War II they drew an imaginary line down the middle of the fjord and forbade the locals from crossing it. Straddling that line now is the gleaming black hull of an enormous cruise ship flying a French flag. Zodiacs are ferrying passengers through the drizzle to the shore from the floating hotel. Jiaggi steers carefully around it and we moor next to a rusty old cargo boat from which a forklift is unloading pallets of Coke, beer, toilet paper, sugar, milk powder and plumbing supplies. Apart from the Inuit, who lived off the land for a time before moving to easier territory across the water in Canada, Greenland has never been self-sufficient. Surviving here is hard, lonely and expensive.

Battered-looking houses teeter above the harbour. In one yard a dog circles on a rope beneath Arctic char drying on a line, the heads of the fish on sticks. Next to another, a tethered horse has grazed a perfect circle in the grass above the seashore. Jiaggi disappears into the cafe that doubles as the post office, leaving me to contemplate the petrol station – a container with two self-service pumps linked to an ATM machine. The tourists are heading for the Norse ruins on the hillside above, now a UNESCO World Heritage Site. The Vikings' little scalloped fields surrounded by stones and recently cropped grass dot this part of the shore. A farmer swaddled in a hat, scarf and oilskins roars past me on a quad bike. And all around the mountains merge with the black clouds to obscure the sky.

The last farm to the south has the largest, greenest fields, product of much hard work to reclaim the soil from the stones and tundra. Bales of hay in light green plastic are stacked in neat rows. A white farmhouse stands next to a blue one and a complex of barns, one side concealed by an unusual and distinctive row of trees. The field rolls in a gentle curve down to the beach, where a car has been left to rot. This was the farm of Brattahlid, the powerful wife of Erik the Red, the first Viking colonist of Greenland, who founded the settlement in AD 982. Nearly a millennium later, this was the place that new explorers came to re-establish sheep farming in Greenland. Otto Friedrikssen was the pioneer, arriving in 1924 with his family, supported and encouraged by the Danish government.

Otto's grandson lives in the blue house now with his Greenlandic wife. Ellen has kind eyes, short, steely hair and stylish glasses. In her black T-shirt and trousers, she looks more like a Scandinavian architect than a sheep farmer. Her house is furnished with the usual accents of Danish modernism: blonde wood and glass. The exception is a table made of what looks like driftwood.

'Greenlandic retro.' She laughs. Before timber was imported, Ellen's ancestors relied on driftwood from Siberia, trees frozen into the ice and then released by icebergs sent on their way south down the eastern coast of Greenland by the Beaufort Gyre, a cyclonic current that, although it is weakening, still spins anticlockwise around the North Pole. The wood, mostly Siberian larch, provided a significant source of lumber for utensils, skis, tent poles and houses. There is less driftwood washed up on the beaches of the fjords now, perhaps because of logging in Siberia, Ellen thinks, or changes to the currents, or the sea ice breaking up further away, releasing the driftwood into the North Atlantic instead of the natural eddies of the fjords. This may actually have been the original reason for the settlement here: trees on the other side of the world sustaining life in this most inhospitable place.

Ellen tells me the story of Kaassassuk, after whom the village of Qassiarsuk was named. Kaassassuk was an orphan who lifted a huge piece of driftwood from the sea while everyone else was asleep at night. In the morning no one could lift it and no one could understand how the driftwood had got so far above the waterline. They realised there must be a very strong person among them, but they didn't know that it was the orphan. There are other stories too, more recent but from before the Americans came: her husband's grandfather told her about going across the water to Narsarsuaq to fetch sticks for firewood from the birch thickets. There was a different kind of forest then. And she tells me how the trees now obscuring the view from her window were part of a planting project using specimens from Alaska. She and her husband put the spruce in front of the window where they could see them forty years ago. They had no idea the trees would end up dwarfing the house.

But it is not the scarcity of timber that worries her now, it is the shortage of rain. I point out the window to the rain battering the fjord and she smiles ruefully. 'Too late.'

This summer advisers from the government came to talk to the

eleven sheep farmers of Qanasiassat Fjord about how to adapt to
a warmer, drier climate. Lack of water severely impacts the hay
harvest and thus the ability of the sheep to get through the long
winter – hence all those carefully husbanded bales. The talk was of
irrigation using meltwater lakes, but even these are drying up in
the summer now. There were only two days of rain this July and
the hay harvest is down 50 per cent.

Ellen first noticed climate changes in 2006. The fjord always
froze in winter. They would drive across to Narsarsuaq. In 2006 it
didn't, and it hasn't been safe to drive on the sea ice since. There
hasn't been any snow at all for the last three winters, and their
snowmobiles sit rusting in the garage. One farm is experimenting
with cows. Ellen doesn't know what to do. The future for sheep
seems bad. They had been planning to pass the farm on to their
son, but he too is less certain about the future now.

Ellen appears calm and resilient. She is also the head teacher of
the village school in Qassiarsuk, which closed for lack of students
in 2014 but has recently reopened with twelve children and three
teachers. It is only when she speaks of culture that her anxiety has
an emotional edge. She shows me her national costume – white
sealskin shorts, an embroidered shirt, red woollen leggings and
white sealskin boots. It is a challenge to keep making these clothes
in the mild winters, she explains. The colour of the sealskin is
achieved by drying it outside in the desiccated cold of winter. If it
is too mild, the skin does not turn white.

'People are starting to make our traditional costume with fabric!' Ellen says with more outrage than she has mustered so far. She
indicates the intricate tops of the boots that look like beading but
are actually tiny strips of dyed sealskin laid tightly together. 'The
people with the skills are dying out – the knowledge. No one
knows the names any more . . .' Her voice tails off, her hands fall
to her sides. It is as if survival is one thing, a technical, modern
challenge that can be weathered, but the death of her culture is
irreparable, unforgivable. This is how the world ends. In a myriad
of tiny tragedies. Each extinction of species, language, custom
noted not with a howl of protest but with a quiet tear.

Once again Greenland is proving itself a laboratory for civilisational collapse. The Norse colony disappeared some time after 1420. The Vikings farmed sheep, cattle and goats and grew hay and wheat with difficulty. At its peak, the colony had a bishop, a cathedral, twelve churches and three hundred farms. They exported walrus ivory and polar bear hides to Norway, and imported timber from the Labrador coast of Canada. When the climate got colder in the Little Ice Age, which began around 1400, summer sea ice apparently prevented their ships from leaving the fjords. The archaeological record suggests that they ran out of wood for fuel; in less than 500 years they had managed to clear all of the low and slow-growing birch and alder thickets that had taken thousands of years to form. But well before all sources of wood had been exhausted, they also took to burning turf as an alternative, an even more disastrous choice since soil takes even longer to form than wood.

Jared Diamond, in his book *Collapse*, describes how a combination of environmental degradation and deforestation contributed to the fatal pressures on the Greenland colony.[4] If the Norse had appreciated the fragility of their environment and managed it properly, things might have been different. Environmental degradation, specifically deforestation, Diamond notes, is central to the collapse of every human civilisation that we know has taken place.

Bluie West One was one of fifteen bases in Greenland the American government publicly acknowledged. There was a sixteenth, Camp Century, a secret base under the ice intended as a store for nuclear warheads close to Russia. Camp Century was a venue for multiple experiments about the dynamics of ice behaviour and spawned much of modern glaciology. One experiment drilled a mile down into the ice until the drill bit brought up soil and leaves. The frozen core lay dormant in a Danish university freezer until 2020 when analysis revealed the oldest ever discovered DNA.[5] Between 450,000 and 800,000 years ago, Greenland was densely wooded with spruce, pine and alder with many insects and beetles. Mean average temperatures were several degrees higher than present.[6] Greenland was undeniably green. And it will be again.

A treeline model constructed by the University of Copenhagen predicts suitable soil and climate zones for tree growth at all latitudes in Greenland up to the northern coast by 2100.[7] This suggests the microclimate of Narsarsuaq, or the Tunulliarfik fjord system more widely, that led to Erik the Red christening the island Greenland could again become a unique climate niche supporting a much denser array of species. For southern Greenland as a whole the model showed climate analogues across much of North America, Scandinavia, Siberia, Scotland, the Alps and even the Carpathians and the Urals. This means that, given adequate soil and water, Greenland could sustain forests like those in Germany or Romania. And, more importantly, it could be a refuge for species facing heat stress much further south.

The problem with most refugia is what scientists call capacity. How long can they serve as refugia for? In most forecasts, the rate of warming unleashed by humans is so fast that areas appropriate for refugia in 2100 will be obsolete within 80–100 years as warming accelerates. Greenland could be different. Even at higher temperatures and with all the associated feedbacks, its ice will still take a long time to melt. Perhaps thousands of years, perhaps hundreds.

There is still three million square kilometres of ice left, some of it several kilometres thick. The proximity of the ice cap and, crucially, its melt waters to a valley system of microclimates could act like a fridge keeping the area comparatively cooler for longer, allowing species a toehold safe from the drought and fires that are predicted for the rest of the boreal zone. The ice sheet may have passed the point of no return, but it has a long tail, and that lasting echo of ice will likely define the composition of the next iteration of the boreal forest for millennia to come.

The next day Kenneth must leave his beloved arboretum for an appointment in Washington DC. He identifies as Greenlandic, tracing his heritage to its admixture of Inuit and Danish colonial blood over several centuries, and with Greenland still a province of Denmark, albeit with its own devolved government, Kenneth

has enjoyed a varied career. Having worked his way up the ranks
of the Danish foreign office as an agricultural adviser to the country's aid agency, DANIDA, in Asia, Kenneth's day job is now
deputy foreign minister of Greenland. In August 2019 Donald
Trump's pitch to buy Greenland is top of the news. It is not the
island's possible role as a life raft for climate change that has moti-
vated the president's interest, but its rich mineral resources,
including uranium, newly accessible due to the melting permafrost. There is also increasing US interest in reopening some of its
former military bases. Of sixteen there is now only one, called
Thule, next to a sacred mountain holding the spirits of generations
of Inuit. A runway overlies the site of the Inuit village of Qanaq,
which was forcibly moved one hundred miles north.

I wake to the sound of rain peppering my tent. The inside is
slick. I am camped at the foot of a scree slope near Kenneth's cabin.
A telegraph pole, a giant of a tree grown on the edge of the Pacific
far away, now lies in a ditch, its wires tangled with willow. The
ground is a carpet of lichen which yesterday crackled like frost
underfoot but is now springy as rubber. Juniper and willowherb
streak up the hillside before petering out at the start of the rock. In
the eerie grey dawn the lonely *clock clock* of ravens echoes off the
towering cliff above, a reminder that there is almost no other
birdsong.

I tiptoe into the cabin to find Kenneth, all in black, packing his
black bag and drinking black coffee at a round table by the window, beneath a polar bear skin nailed to the wall. He doesn't like
to talk about politics, but he can't resist letting slip his destination:
the West Wing of the White House! But when pressed about
Trump, he becomes taciturn again, his grey-green eyes narrowing
with the calculation of the diplomat.

'Let's just stick to the trees.'

Whatever secret dealings he is party to in his government job, it
is probably the trees that will be his most enduring legacy. Together
with Poul Bjerge and Søren Ødum, Kenneth has almost certainly
changed the vegetational and geological history of Greenland.
And, if Greenland's potential as a critical refugium for boreal

species is demonstrated further, he might well turn out to have played a key role in determining the shape of the global forest that will emerge after our current moment of environmental vandalism has passed. At some point, human emissions will slow and stop. When all of the methane and carbon frozen in the permafrost has been released, when the feedbacks have played themselves out, and the mix of oxygen and carbon dioxide has stabilised again, whatever photosynthetic machinery is still surviving will begin once more the painstaking process of building a forest with the resources of soil and seed available. Here in southern Greenland the combined efforts of Kenneth and the American military will determine what those resources are likely to be.

Down at the seashore, sheets of fine rain drag across the valley floor. A muddy stripe lies across the fjord. The roaring creamy river is cut with brown soil. Within the fence of the airport runway two large snowshoe hares lollop over the scrubby gravel. Kenneth's diffident sandy head disappears through the double doors of the terminal just as the former glaciologists Dirk and Fazia and their son Radin arrive with Peter the forester, back from the south. I join them, jumping in the back of their red truck to head down to the site of the old airbase to plant some more trees.

From the airstrip, the old American road curves over a bridge and onto a flat plain dotted with hundreds of small trees: Siberian larch and Engleman spruce. This is the trees' favourite place, the area where they most freely self-seed. Unlike the fjord with its sedge and moss and nowhere for the conifers to establish, this disturbed glacial moraine, levelled by American bulldozers in the 1940s, is ideal soil for seedlings to take root. We take the bazookas and a shovel and search for gaps within the emerging forest where some trees are over two metres high. Amid the broken concrete, drainage channels, felled steel pylons and collapsed telegraph poles left by the Americans, shoots are springing up. Everywhere we dig, we unearth rusting bits of machinery, rivets from aircraft, forged tubes, plastic pipes and metres and metres of wire. In one spot a whole spray of larch has self-seeded in the crumbling remains of a bitumen felt roof. In another a slab of asbestos is serving as a nursery for spruce. Once, Kenneth excavated a crate of 1940s Coca-Cola bottles. The oddest spectacle that morning is a cast-iron fire hydrant still standing among the trees. It is a portent, a snapshot of the future. Where once was a miniature city with a hospital, a school for sixty children, four permanent orchestras, seven clubs and five thousand servicemen for whom Marlene Dietrich sang in 1944, there is now, less than seventy years later, a nascent forest.

The Americans would drink cocktails made with ice chipped from the bergs floating in the fjord. The ice, formed at intense pressure over thousands of years, would fizz, crackle and pop in the glass. But there are few bergs now, and the snout of the glacier is no longer visible from the site of the base. It has retreated hundreds of metres up the head of the valley behind a bluff of rock.

Still, the brooding presence of the ice is everywhere: in the constant roar of the river, fed by the melting glacier, the shape of the valley itself, and above all in the light, the lambent glow from behind the hills that inverts the sky, making the horizon lighter at the edges where the arc of the atmosphere touches the ice cap and is refracted back. The ice is Greenland's heaven and hell, its constant, totalising force, giver of life and death. Dirk and Fazia remember the glacier being much closer when they first came here. And Peter recalls walking on it several decades ago. They all insist that a visit to Greenland is not complete without seeing it.

The treeline has evolved in dialogue with the ice, always maintaining a respectful distance. All across the treeline ecotone it is the ice that dictates which trees can grow where. Orogeny, the combination of tectonic movement, glaciation and erosion, shapes the geology of the earth's crust. As the glaciers retreat, the resultant shifts in catchment, drainage and minerals change soil composition, nutrient levels and drive the adaptation of plants, trees and associated life forms. The fate of the forest and the ice are entwined in a planetary tango which lasts thousands of years, although the current dance looks to be the last one for quite some time. According to one study, a tiny increase in carbon dioxide in the eighteenth century at the beginning of the industrial era, from 180 to 240

parts per million, stopped the next ice age from being triggered.[8]
Other studies have suggested that the next ice age might have been
due in 23,000 years but now looks as if it will be postponed, if not
cancelled altogether.

I set off for the ice cap just after dawn, as the rose-gold glow of the
sun first pricks the needles of the peaks to the north. On the American tarmac road that leads inland to the head of the valley the
dawn light steadily slides along the ridge and down the opposite
side of the valley, picking out the yellow, amber and pale evergreen
of the birch, juniper and willow underbrush that has staked its
claim on the steep hillside: the forest's leading edge. As the road
disintegrates into a track of granite boulders, I am on the lookout
for the distinctive slim symmetrical leaf of the rowan. In a week
here, the only mountain ash I have seen has been a *Sorbus aucuparia* from Sweden in the arboretum and other specimens from
Iceland planted around Kenneth's cabin. Kenneth said the Greenlandic version can be found in the mountains – the mountains that
hold the ice at bay, towering above the little valley, making its flat-
ness notable and unique, worthy of naming. In Inuit *narsarsuaq*
means 'the flat place'.

The English 'rowan' is a direct borrowing from the Norse word
raun, meaning tree, and in Scotland it is still pronounced that way.
For the Norse it must have been the most important tree and for
some their only one. In their mythology, man was formed from an
ash tree and woman from a rowan. In the Celtic ogham script the
rowan is *luis*, and in Welsh it is *criafol*, the 'crying tree'. In old English rowan was known as *cwicbeam*; literally, 'live tree' or possibly
'tree of life'.[9] To the Celts the rowan was a portal between worlds,
the threshold of the spirit world, invoked to conjure spirits or
inspiration.

In the Inuit language, the word for tree is *orpik*, which is also
the word for birch. It makes sense. Beyond the ruins of the American
base the ground is thick with the hybrid birch, *Betula glandulosa*.
Their oversized leaves are tinged yellow and pale orange; August
in Greenland is the beginning of autumn. You can tell the species

in the arboretum imported from more northerly latitudes because they are a deeper shade of red – their genetic memory anticipates shorter days and so they shed their leaves earlier.

The valley narrows into a gully where the birch are about as high as my head. Underneath, willow and juniper cover the rocks. The sun has cleared the mountains ahead, and over the rim of the valley pours the golden dawn, setting the rocks alight, but I feel colder. It takes me a while to realise why: I am approaching the ice.

The breeze flowing off the glacier is bitter, pinching the skin on my face. The air is so clean and crisp it could cut paper. The smell of water is mixed with the scent of thyme and juniper crushed beneath my boots. At the crest of the pass, before the path descends into a valley of grass and flowers leading up to the glacier, the birch are down at my knee within a matter of metres. The ecotone, the ecological transition zone, is foreshortened by altitude but also by the presence of the ice. The treeline zone, which can stretch over hundreds of miles in Canada or Siberia, is compressed into a matter of metres in Greenland. The valley only steps behind me is subarctic, boreal; the escarpment ahead rising above the snout of the glacier to the desert of the ice cap above is unmistakably tundra: the green, fawn and purple of low grasses, alpine plants, mosses and lichens: cottongrass, angelica and willowherb.

As I descend into a crook of floodplain, the wind drops, the temperature rises somewhat, the trees gain in height once more and suddenly there is a sound that had been missing all morning: birdsong. Not much, but a comfortingly familiar *cheep chirp* from a trio of small brown birds that resemble stonechats. Kenneth said that the warmer temperatures have brought more vagrants to the arboretum. The Bohemian waxwing (*Oleanthe oleanthe*) is now a common visitor to Narsarsuaq. Perhaps this explains the dearth of Greenland mountain ash among the low birch forest of the upper valley. A mountain ash seed must pass through the acidic stomach of a bird before it stands a chance of germinating. Could it be that the disequilibrium dynamic that limits the spread of the species is the lack of seed-eating birds in Greenland while the wind-pollinated birch is clearly having a field day? If so, global heating will help as more vagrants find their way north.

The brilliant sun on the pinkish cliffs and the starched blue of the sky, which has been mostly hidden all week, make the morning sing. The scent of a meadow is so heady it should be bottled. The hay has been freshly cut: huge plastic-covered bales guide the eye to a combine harvester abandoned mid-job, its windows covered in sparkling dew. Beyond, the path crosses the meadow to a wide bend that the flooding river has worked into a series of interlinked channels. The little bridges have been overwhelmed and carefully placed stepping-stones lie visible in the clear stream, half a metre underwater. Feet have cut a higher path along the edge of the valley, around drowned shrubs, riparian willow now floating midstream. The roar of the main river is all around. Grey water cradling slabs of dirty ice meanders around a cliff and then widens into a foaming skirt over even-sized white granite boulders that snag the ice and make it dance and nod until it falls apart and joins the sea-ward torrent.

The apron of the glacier is a wide floodplain covered in banks of stones all sorted according to size, a product of the wondrous workings of ice. The weight of a moving glacier will concentrate stones like ball bearings into perfect gradations of diameter. As I approach, the grass gives way to moss and alpine cinquefoil – *inneruulaaraq*, 'the little one that looks like fire' – and finally to pale red and green lichen, then nothing but bare rocks and sand. The snout is hidden behind a shoulder of cliff, its scale suggested only by the mass of swirling grey water surging between the vertical walls of bare rock that frame the valley soaring hundreds of metres into the air.

The sound of the river is deafening and feels loaded with meaning. It is no longer possible, if it ever was, to sense or describe an environment in neutral terms. It is impossible to hear only a rushing river. It is impossible to unknow where the water is coming from, why it is arriving in such volume and with such force. Impossible to silence the unbidden question: what does the river say? It is the sound of guilt, the sound of reproach, of fear.

To actually see the ice these days, one must trek up a treacher-
ous path with the aid of ropes onto the plateau above. I climb.
After sixty metres or so, I see the last willow and the last juniper
flattened to the ground alongside the path, about the size of a
human arm. Another sixty metres of climbing, and I am high above
the valley looking back to the inky sea and turquoise sky and the
jagged black and white of the stunning vertical peaks, giddy above
the neighbouring fjord. Ahead, the plateau is dotted with crystal
lakes, deep, cold, devoid of life. Sedges and grass form a thin covering over the rocks. The path weaves its way between huge
boulders. Four Canada geese are sheltering beneath one and erupt
squawking almost from under my feet, landing a minute later on
the mirrored lake below, making the sky ripple in response.

Then the path dips to a former melt lake that is now an
evaporated basin of thin mud overrun with Arctic cottongrass
(*Eriophorum scheuchzeri*) – *ukaliusaq*, 'the one that resembles a
hare'. Its wispy flowers surrounded by fine fibres permit the plant
to capture light and increase the temperature of its internal repro-
ductive organs, allowing it to convert nutrients faster than other
neighbouring plants. The Inuit used to eat the stems and use it to
treat diarrhoea. It will pave the way for other species, of which at
present there are precious few.

Over the final ridge, the full majesty of the ice cap is suddenly
revealed. A blinding dirty white sea with black pyramids poking
out – mountains up to their collars in ice. In the foreground the
plateau falls away to a steep edge above the deeply cracked and
fissured surface of the minor yet still magnificent glacier named
Kiattuut Sermiat. Peter said that formerly one could climb *up* onto
the ice, but there is now a drop of several hundred metres to its
grubby surface, streaked with black. This is the new skin disease
afflicting the surface of glaciers: soot from far away and black carbon and algae feeding on organic nutrients released by the melt. A
darker surface absorbs more light from the sun and begets more
melt, another of the cruel feedback loops accelerating warming in
the Arctic system. The ice is a negative image of the sea behind,
resembling a finger of ocean narrowing to the mouth of a fjord,

stretching away to a wide vista of inverted frozen ocean, ridges
and troughs of waves flecked with dark spume.

Between the rock and the ice is a grey pool fed by a steady
trickle pouring off the surface of the glacier between furrowed
ridges. The pool tinkles with dissolving ice. At one end the cloudy
soup disappears into a moulin – a black void beneath the glacier –
the submerged rush of the water making a strangely suffocated
sound. The grey stream falls away into the dark, leaving only an
echo and the eerie, incessant, deadly trickle. There is no sound
apart from the melt, filling the immense silence of the mountains
and the ice cap with the force of a scream.

It has taken nearly half the day to get here and it will take
another half to get down. I need to move soon if I don't want to be
stranded on the plateau in the freezing dark. But I cannot tear
myself away. The mesmerising immensity of it. Its mysterious gift
of life and the promise of oblivion. People come here to study its
secrets, to mine its wisdom, to consult the archive of the planet's
climate and to see it before it is gone. Perhaps this is what accounts
for the fascination of so many cultures with the cold extremes of
the earth, the explorers and writers and tellers of stories of myths
and fairy tales, of snow queens and Narnias. Deep down, we know
we need the ice. As long as humans have been here, so has the ice.
The fascination also derives from its power and, by contrast, our
own powerlessness. It is separate, ungraspable, uncontrollable. Its
crystalline perfection holds not only all human history but our
future too. And we sit by and watch as glaciers spill our past and
our future faster and faster into the warm bath of the sea. To face
the ice is to contemplate death.

The tango that the ice and the trees perform together has cooled
the planet over millions of years. Without trees to reduce the carbon dioxide in the atmosphere, the ice might never have been able
to form in the first place. It was a reduction in carbon dioxide that
created the tipping point for the Quaternary glaciation which
wiped out billions of plants in order to allow them to be born
again through the 100,000-year pulses of the ice ages. We never
knew the Eden of the last 10,000 years was so fragile. Without this

delicate heartbeat of ice, the earth might never have evolved the strange equilibrium of the Holocene, which has witnessed such an extraordinary profusion of biodiverse life on earth.

The planet is a finely tuned system. A few degrees of change in its orbit can usher in an ice age; a few degrees of temperature change can transform the distribution of species, can melt glaciers and create oceans. In the future, when the ice is gone, there may be no such thing as a treeline at all. As the stable currents of air and water associated with the Gulf Stream, the polar front, polar vortex and Beaufort Gyre dissipate or fluctuate, the Arctic Ocean melts completely, and the Rossby waves in the upper atmosphere go haywire, the fine gradations of temperature, altitude and latitude first observed by Alexander von Humboldt will become decoupled and ecological transition zones scrambled. Instead of a majestic sweeping zone of forest around the planet, we might find discontinuous pockets of trees in odd places, refugees from soil and temperatures long gone, and crocodiles once again at the North Pole.

Global heating messes with the most essential functions of the earth that we and other species need to survive: the cycles of breathing, the pulses of life. Not just the geological relationship of ice and treeline, up and down in tandem over hundreds of thousands of years, but the annual photosynthetic pulses of seasonal production – the spike of oxygen in spring when the trees put out their leaves, and the daily peaks and troughs between night and day that regulate the primary chloroplastic functions of the plant kingdom. These pulses are literally the heartbeat of the planet – they oxygenate our world – but the peaks and troughs are getting shallower, less well defined, as the oxygen fraction trends downwards. When there is more carbon dioxide in the atmosphere, trees don't need to work so hard to get their daily fix of carbon. They conserve energy, they open less of the stomata on their leaves. This means they inhale less, they transpire less, they breathe out less oxygen.[10]

Suddenly I feel short of breath. Under the midday sun, the sheets of white flash brilliantly, blinding and disorienting. I have been

standing on the cliff edge for too long. I start to feel dizzy, nauseous, my limbs light with fear. It is akin to the prelude of a panic
attack, as if I have just had a brush with death. Which, in a way, I
have. Research by Jason Box, the American glaciologist I met at
the tree planting, will reveal that during the 2019 season the
Greenland ice sheet lost 254 billion tonnes of ice, seven times as
much as the 1990s. Ice sheet melt on this scale had not been
expected until 2070. And it is accelerating. What's more, in 2019
the melt season continued into October, and in 2020 into December, setting the stage for rapid ice sheet collapse.

We know what is happening. An unfortunate side effect of science is the illusion of human mastery: the idea that if we know
what is happening then we can control it. The irony is that we
might have been able to. The tragedy is that it is too late. The chain
reaction is under way. The curve only gets steeper from here. From
emissions already in the atmosphere, Jason says, five metres of sea
level rise is locked in; it's just a question of how fast the ice melts.
Once again, the models seem to have underestimated the speed.[11]
All those stories and ideas of the frozen north, like so much else of
human culture linked to stable climates, familiar species and regular seasons, will be like the light from stars shining for years even
when they are long dead.

Under my feet the reddish-brown rock is scoured with deep
grooves, striations from the glacier's recently completed geological
work. I have a strange feeling of time telescoping. I am standing on
a geological fault line, the beginning of the Arctic–temperate transition zone, the very edge of a process that finished almost yesterday
in planetary terms, when Wales, Scandinavia, Siberia, Alaska and
the Canadian Shield were covered in ice, and in the blink of an eye
the treeline has chased the melting ice north. It is amazing, actually, how fast nature moves.

But in the next blink, when Greenland is forested, when the tree
island of Taimyr is no longer an island, when there is no tundra left
in Scandinavia or Alaska and the forests of North America and
Siberia are burned-out prairie, when the trees are where I am
standing now, will there still be humans to see them? Our current

epoch brings to mind a question posed by Bishop George Berkeley in the seventeenth century but then gives it a twist: if a tree grows in the forest and there is no one to see it, did it really happen? Is it possible for humans to imagine a planet without themselves?

Our present emergency is forcing us to remember what, until recently, we have always known: that there is a web of communication, meaning and significance beyond us, a world of life forms constantly chattering, shouting and flirting and hunting each other, indifferent to human affairs. And there is solace in such a vision. The way out of the depression and grief and guilt of the carbon cul-de-sac we have driven down is to contemplate the world without us. To know that the earth, that life, will continue its evolutionary journey in all its mystery and wonder. To widen our idea of time, and of ourselves. If we see ourselves as part of a larger whole, then it is the complete picture that is beautiful, worthy of meaning and respect, worth perhaps dying for, safe in the knowledge that life is not the opposite of death but a circle, as the forest teaches us, a continuum.

The faint *pitter-patter* of a helicopter breaks the moment, a small red box carrying scientists or tourists up onto the ice sheet for a glimpse of eternity on the move. I set off downhill, awed, moved, made small. Coming to terms with the earth's immense, priceless and vanishing trove of ice has left me spent, and my legs wobble down the path. The sun follows me down. The mountains are sharp and beautiful in the afternoon light. The sea in the fjord bounces light up the valley, ravishing in its autumn coat of gold, red and green. The red, grooved rock gives way to gravel, the path loops up and then down over the moraine in front of the retreating glacier snout. I am no longer walking on fresh ground revealed to the air after thousands of years trapped under the ice; my boots are springing over moss and grass finding a home amid the gravel, silt and rock. Lichen is starting to mine the boulders for the beginnings of that precious soil. Ahead there is willowherb, juniper, birch, and beyond are the noble spines of the conifers, their winged seeds rising on the wind. The forest will be here soon.

Epilogue: Thinking Like a Forest

Llanelieu, Wales
52° 00' 01' N

A forest is not a static thing; it is a constantly evolving mosaic of species in multiple relations with each other as well as with the rocks, the atmosphere and the climate. The pioneering Russian ecologist Sukachev called this interrelated system biogeocoenosis. The Koyukon call it 'the world that Raven made'. The precise workings of this most complex of relationships are a mystery whose contours we can only guess and whose outcome we can only admire in the shape of the living breathing forests that sustain life on earth.

This book has been an attempt to catch a glimpse of nature's algorithms at work and to stop and wonder at the results. It has not sought to offer any solutions to the crisis in nature that humans have wrought although some conclusions are inescapable. There is much to fear in what we know and much to hope for in what we do not.

What is plain from my journey along the treeline is that global warming is well advanced and that while humans may still be able to temper the scale and severity of the warming unleashed, they are powerless to stop it. Also, that even during the short time that I have been researching this book (2018–21) the speed of the geological changes being witnessed is accelerating faster than models predicted. The world is in the grip of unprecedented change. The planet you think you live on no longer exists. But this is old news.

The real question now is what we do with this knowledge. Accepting the reality of our rapidly transforming environment is a fundamental challenge to affluent ways of life and to Western habits of mind founded on notions (and experience) of progress, peace, democracy and economic growth. Conversations in the global north seem trapped between increasingly unrealistic dreams of 'net zero' and painless green growth on the one hand and misanthropic stories of apocalypse, violence and human extinction on the other. But the history of the people living along the treeline – who have been living with the reality of a transforming environment for longer than most – offers an alternative. There is a third story, a more positive reading of humans' relationships with their habitat that holds the key to imagining a different future.

Trees and humans enjoy the same climate niche. Our opposable thumbs are a constant reminder that we evolved, and thrived, in trees. We will always be creatures of the forest. For the 11,000 years since the last ice age humans have co-evolved with trees, moving into the habitats pioneered by the advancing treeline and then adapting, managing and stewarding them, engineering a remarkably stable and convivial environment for nearly all that period on a global scale. The Koyukon, the Sámi, the Nganasan, the Anishinaabe are just a few of the countless indigenous peoples whose world view attests to our foundational reliance on the forest. We have been the keystone species of the Holocene – a geological force, certainly, and not an entirely negative one. There is barely a forest on the planet that has not been disturbed by humans, often opening niches for biodiversity.

The Zimovs are almost certainly right in suggesting that *Homo sapiens* exterminated the megafauna of Siberia, paving the way for the taiga forest, but we also carried the Scots pine to Scotland and balsam poplars to the gravel eskers on the shores of Hudson Bay. We coppiced the cryolithic larch forests of Ary Mas as well as the temperate rainforests of Wales and Scotland, pastured animals in the woods, carved out hay meadows, fens, levels and moors and managed the burning of the boreal like the Anishinaabe of Poplar

River precisely to promote biodiversity for human benefit rather than destroy it. Our short reign as the planet's keystone species has coincided with the summit of biodiversity on earth. As the radical ecologist Ian Rappel has written, 'From the perspective of bio-diversity and the biosphere – there is nothing wrong with the Anthropocene; what is wrong is the . . . way it is now being run.'[1]

The breaching of the ecological ceiling of the planet was only enabled and accelerated by a specific recent economic model: industrial capitalism and its political export, colonialism.[2] But capitalism, the exploitation of resources and labour to concentrate wealth in the hands of the few, is not necessarily the best economic model; indeed our collective survival on the planet almost cer-tainly depends on moving beyond it. Surveying the territory from within our capitalist moment we are encouraged to believe there is no alternative and that the crisis is our fault. But accepting the blame is not only profoundly disempowering, it is also misplaced.

We did not choose our economic system from a list of all avail-able options. We are all to a greater or lesser extent the victims of historical forces that have built structures of power over centuries based on a very blunt assessment of value. For a tree, only the tim-ber fetches a price at market, not the soil that grew the tree or the insects that pollinated it, the sun that fed it or the rain that watered it. But the community of forest that is home to so many species has no price. Capitalism not only alienates and commodifies nature into products and transforms humans into consumers, it also alien-ates and commodifies us. Our very gaze has become a product. Our attention is directed away from the biosphere that sustains us, and this alienation makes us to varying degrees blind, deaf and dumb. Looked at in the long history of our co-evolution with the forest, the human break with nature is an eye-blink; the story of human life on earth is longer and wider than the history of capit-alism and, most importantly, the ending is not written yet.

We are not born indifferent to our surroundings. As I write this the whine of chainsaws is echoing through the mixed woodland of the small narrow valley called Cwm Rhyd Ellyw that falls away from the church of Llanelieu next to my home. It is what's called a plantation on ancient woodland, and so, despite the presence of some mighty hardwoods, the authorities granted the landowner a clear-fell licence to raze the lot. All the characters of the boreal were there – Scots pine, birch, larch, spruce, rowan – and several others, like alder, ash and Douglas fir too.

When I walked down the road with my two daughters to survey the damage done to the places they used to play, the stone shaded by a birch tree they called the river cafe, or the deep pool below a fallen poplar known as Heron's home, they were so shocked they burst into tears. Logs were piled high, the air heady with sap, the steep stream banks littered with branches and the rain filling the deep tank tracks of the logging machines making the river run a cedar shade of red. The girls could see the mountains beyond for the first time in their short lives. They are six and four. They asked why the trees were being cut down when we need the oxygen they produce and the rain they condense, but what upset them the most, causing another round of sobs, was concern for the living things that call the forest home.

'The trees must have been crying!' (An idea familiar to the First Nations of Canada.)

'What if a mother ladybird came back to her nest only to find the tree had been cut down and her babies were gone!'

Parenting in an era of rapid climate change does not allow for the luxury of misanthropy or false hope. We must, to paraphrase the University of California anthropologist Donna Harraway, 'stay with the trouble'.[3]

My imagination was gripped by the moving treeline after finishing my last book about refugees in the Horn of Africa. This was not just the desire to go somewhere cold after the unforgiving heat and dust of the equatorial desert of Kenya and Somalia. The Horn of Africa, like the Sahel belt in general, is particularly sensitive to climatic changes in the oceans and rainfall patterns driven by teleconnections with forests – and deforestation – elsewhere. The displacement and violence there are largely driven by drought and climate change, and I wanted to write about other places where the impacts of warming were already visible, where one could catch a glimpse of the future.

I did not realise how relevant my experience of reporting on war and refugees in Africa, about people struggling to find meaning and hope in difficult circumstances, would turn out to be. Victims of wars or natural disasters are often far better equipped to imagine and respond to dramatic change. In a disaster the social order is stripped away and we are revealed to ourselves anew. The 'human' is unleashed, freed of its habitual constraints, sometimes with barbaric consequences but more often with positive effects. People are capable of extraordinary things. What I learned in the ruins and refugee camps of Congo, Sudan, Uganda and Somalia is that struggle produces hope, not the other way round. Hope is not an inert precious metal lying around waiting to be discovered, it is something that must be manufactured and redefined in the light of shifting circumstances, every day. And here the lesson is that despair is the first step towards repair. Acknowledging the damage of the past is empowering, like the elders of Poplar River transforming colonial grief into a movement that secured North America's largest protected forest, or Thomas MacDonnell reversing centuries of over-browsing by sheep and deer to begin the restoration of Scotland's great wood.

It is a bourgeois conceit that hope has become synonymous with saving – or attaining – an ideal state of affluent security, especially when such affluence is incompatible with planetary limits to economic growth. Māret Buljo in Norway would laugh at such an idea. Hope lies in shared endeavour, in transformation, in meaningful work for the common good.

We are on the brink of a new epoch in the life of the planet. At least two degrees of warming is already 'committed' – in the pipeline – although some scientists project even more than that, up to four degrees of 'implicit warming'.[4] Before the end of the twenty-first century there will be a wave of extinctions, trees will leap north, steppes will expand, the tundra will disappear along with the Arctic sea ice, the oceans will be reconfigured and cities will flood. The last generation to know a stable climate with seasonal cycles and familiar species – and all the human culture and traditions that rest on that foundation – has already been born.

This is hard. But accepting that the status quo is irretrievable is also the door to action. Suddenly there is so much to do. The struggle to limit the damage and prepare for what is coming has already begun. This is the philosophy underpinning Black Mountains College, a new-model educational institution which I co-founded and which is informed by the same research that led to this book. The philosophy starts from a simple insight of the forest school movement: that reconnection to nature will only come when nature itself is the classroom. Teaching outdoors, the college provides pathways for learning the skills and mindsets necessary for organising human society along ecological principles of diversity, balance, limits and symbiosis. If how the treeline made our world habitable in the first place, if how forests create rain, drive winds, manage water, seed the oceans, provide the foundations of much modern medicine, cleanse the air of man-made pollution and disinfect the atmosphere were more widely taught and understood, it would be much harder to cut them down.

More than any previous generation, the lives of children born this century will depend on how it goes for the non-human world. When water and food become scarce, as they are already showing signs of doing, when intercontinental supply chains are no longer viable, when industrial agriculture falters, we will need to once again pay attention. We will need to rejoin the forest, to reverse what Charles Eisenstein calls 'the story of separation'. And the way that we do that is by using the portal through which our souls relate to the rest of the world: through our senses. Curiosity and noticing are the humble but radical prerequisites for a new relationship with the earth. Systems change when there is a culture that demands it. The revolution begins with a walk in the woods. How did it ever come to pass that we would forget the names of the living things that manufactured oxygen, purified the air and water?

If we want to be part of the assemblage of species that co-evolves to survive the coming upheaval then we need to revive that essential entanglement with other living things. We all need to learn once again how to think like a forest.

The Koyukon hunter in search of a bear will not name the bear. He will even avoid looking at the bear. The Nganasan storyteller relating a story about a woman in a *chum* (tent) will not say the head of the household's name, she will be referred to only as 'the one who sits by the door of the *chum*'. It is a habit present in other indigenous cultures and oral traditions too – not to use people's names, but to refer to them in terms of who they are in relation to the speaker: 'Hey, brother!', 'dear sister-in-law', 'teacher', 'older-than-me'. This is a neat acknowledgement that all existence is relational. Each person is not reducible to a single self, but contains many selves, many instances of their self; every creature contains possibilities.

To name the bear is to objectify the animal, and so is offensive. We do not know how the bear refers to itself, nor do we know what other souls or selves the bear's body might contain. And so not naming the bear is a mark of humility and of respect. Not naming is also an acknowledgement of indeterminacy – that the bear's nature is not fixed yet – its relationship to the hunter is still in the process of being determined, a process that will be decided by how the hunter, and the bear, act. The hunter will avoid looking at the bear too, because to see and be seen is a relational activity. When you smell something, tiny particles of the thing you are smelling are dissolving inside your nose: the thing being smelled is becoming you. Indigenous understandings of sensory perception (and those of modern phenomenology as David Abram so eloquently reminds us) take this scientific fact further: all perception is participation.[5] If you see the bear, then it has seen you, and you are both changed by that fact.

The strict rules and rituals in indigenous forest cultures surrounding how animals and plants must be looked at, talked about, treated, killed and eaten derive from the fact that human survival is deeply entangled with that of other species. When we eat the bear, we become the bear in so far as the assemblage of species that gave life to its body are reassembled in ours. In the light of modern scientific studies of the digestive tract, such views are not as far-fetched as they might at first seem. When more than half of your roots are inhabited by another organism or when you depend on flying insects to pollinate your flowers, you are engaging in collaborative survival. All evolution is co-evolution.[6]

The fearsome taboos of the Koyukon or Anishinaabe, the Sámi or Nganasan are not just a recognition of man's reliance on natural processes, they are also an acknowledgement of the awesome responsibility which rests on the shoulders of the planet's keystone species. Warming is a given, but how species respond to it is a story still being told and one in which humans have a key role to play. Strategic ecology will soon become a core element of national security and community resilience. Assisted migration – helping species move and adapt – will become a key goal of conservation. We are Noah with his ark. We have the power to choose at least some of what survives. The trees we select to survive will determine whole forests and ecosystems – as well as the assemblages of species that rely on them – for millennia. The Anthropocene is only just beginning, and its echoes will dominate the earth even after we are gone.

It has always been the case that life is a moral endeavour, the very act of living a legacy. To look at the forest through Celtic, Koyukon, Sámi, Nganasan or Anishinaabe eyes is to see a world of multiple selves and souls communicating with each other. If we acknowledge all this other life and our dependence on it, we have to confront the question: what is the right thing to do? The leaf talks to the wind, the flower talks to the bee, the roots talk to the fungi – the world is a chaotic, noisy place! When we step into the forest we are making the world with our bodies, with our feet, our eyes, our breath, our imaginations. A million randomised branching futures are possible. The forest is a sea of possibility, an infinite experiment in co-evolution.

A hopeful future in such a definition is not a prayer for stability, of stasis, but an invitation to participate: to explore, experience, to get lost or find your way. It is an opportunity to realise your true self by doing the right thing. What you have yet to do will always define you more than what you have done in the past. Things cannot be named because they are unfinished. Evolutionary nature is an engine of mystery. Of things we do not and cannot know. In the forest you are part of something magical and huge where every step is simultaneously an act of destruction and of creation, of life. There is consolation in the fact that we have always lived in the ruins of what went before and we are living there still.

We must prepare our children for uncertainty but not as victims. We and they are stewards, still charged with an ancient responsibility. The earth is alive and enchanted, and to act within it is to enchant by living – to see, hear, feel, dance – to create the future with every step in full recognition of the fact that every move you make, however large or small, matters.

Glossary of Trees

The following pages are descriptions of the trees featured in this book – the protagonists of the chapters as well as several other common species of the boreal. These are summaries of what I have learned from others more expert than me. For more information please consult:

Diana Beresford-Kroeger, *Arboretum Americana* (Michigan University Press, 2003)

Diana Beresford-Kroeger, *Arboretum Borealis* (Michigan University Press, 2010)

Daniel Moerman, *Native American Ethnobotany* (Timber Press, 1998)

Nature Conservancy (nature.org)

Iain J. Davidson-Hunt, Nathan Deutsch and Andrew M. Miller, *Pimachiowin Aki Cultural Landscape Atlas* (Pimachiowin Aki Corporation, Winnipeg, 2012)

Trees for Life (treesforlife.org.uk)

Colin Tudge, *The Secret Life of Trees: How They Live and Why They Matter* (Allen Lane, 2005)

Woodland Trust (woodlandtrust.org.uk)

Alnus, alder

Alnus glutinosa, also known as common alder or black alder, is a common sight in the Eurasian boreal forest up to elevations of around 500 metres. In North America many other of the thirty species of alder flourish. They are medicinal trees for the First Nations known as 'the willow that smells'.

Alnus is a member of the birch family, Betulaceae. It is a water-loving species, often found near rivers and lakes, where its deep roots help to stabilise riverbanks and improve soil fertility through its ability to fix nitrogen from the air. The alder grows fast, like the birch. It puts out buds and almost round leaves with toothed edges from multiple stems and also directly from the stump. These buds and twigs can be sticky, hence *glutinosa*.

The tree is monoecious with male catkins and female cone-like flowers which emerge in spring before the leaves. It is pollinated by wind. The seeds have air membranes and germinate on the surface of water, allowing them to take root in riverbanks where they are washed up.

Alder is a critical species for biodiversity, providing food for over 140 insect species, leaves that decompose in rivers, chemicals that protect aquatic life and roots that are home to forty-seven different mycorrhizal fungi. It also hosts a nitrogen-fixing bacterium called *Frankia alni* that takes carbon from the tree in exchange for nitrogen. This makes alder a perfect choice for reclaiming and rehabilitating degraded landscapes, a pioneer fertilising soils with nitrogen for other trees to follow. It was used to reclaim coal spores in the USA and to re-forest the green belt around St Petersburg in 1915.

In Celtic folklore the alder was *fearn* in the ogham script, and was associated with hiding and secrecy. Alder woods – carrs – were wet, boggy, inaccessible places. The wood survives extremely well when immersed in water and so was a common choice for lock gates and canal construction. It was also used for the wooden posts supporting Venice. It is very resistant to fire and is planted as a firebreak in forests.

Alnus glutinosa, common alder

Betula, birch

The birch family is wide and broad and many of the sixty global species can be found across the boreal – from the treeline all the way south into temperate forests. It is the pioneer species par excellence, its abundant seed spread on the wind germinating in a rapid four weeks without much soil at all. The unfussy birch prefers acidic, recently cleared or burned areas where, once established, it provides shade for the saplings of other longer-lived trees of the forest like oak, pine and cedar.

Its distinctive white bark – so flat and fine it was used for paper by some cultures – is more pronounced in some species than others. Downy and dwarf birch (*Betula pubescens* and *Betula nana*) have thicker, more fissured, grey bark, suited to the cold. The leaves of the dwarf and downy species are rounder with a single row of teeth. Most birch leaves have domatal hairs on the underside, an important habitat for insects like aphids, which are food for birds, caterpillars and butterflies. Birch is host to many butterflies and over 330 species of insects overall. Its fungal relationships are equally rich, with a multitude of mycorrhizal partners, and several common mushrooms like chanterelles, the birch bolete and fly agaric depend on it.

Tapping of the trees in spring to obtain its nutritious sap was an ancient practice of many cultures. In Poplar River elders remember *oh-chi-kah-wah-pi* – making containers from the bark to drink birch sap, and *no-skwa-so-wach* – cutting the sweet inner cambium to eat.

Its sap, fungal relationships and the ability to repair degraded soil are some of the many reasons to encourage planting of birch now. The aerosols and resins of the tree benefit humans and animals nearby alike. The leaves are antiseptic, the bark can inhibit tooth decay and the steeped leaves can treat urinary infections.

There is a Native American Odawa legend about the man who became the first birch tree and an Ojibwe story about how the birch tree got its burns. Among Chippewa communities birch bark was used to wrap the bodies of the dead. The *Pimachiowin Aki Atlas* talks of Ojibwe using birch bark to make maps of long-distance routes.

For the Celts, the birch was a symbol of renewal and purification. *Beithe* was the first letter of the ogham script. The Celtic festival of Samhain – Halloween – was celebrated with birch besoms (brooms) sweeping out the old year, and at Beltane – the spring festival – fires of birch and oak were lit. Sometimes known as the lady of the woods, the tree was associated with fertility, hence the alternative to a church ceremony of a 'besom wedding'. The Anglo-Saxon goddess of spring was Eostre, who was also celebrated through the birch and maypoles were made from the tree.

Betula pendula, silver birch

Corylus, hazel

Hazel is the marker of the human relationship to the forest that once was. From the hedgerows of the British Isles to the archaeological record of Mesolithic life in Europe and North America, it is clear that humans relied on the nuts of the hazel tree to a large extent. Hazel represented as much as 75 per cent of the European canopy at one time, leading to speculation that humans were responsible for deliberately spreading the tree. Savannah systems of burning, pasture and cropping lend themselves to nut cultivation, and nuts are a rich source of protein to rival meat.

Hazel is fast growing with many stems – rods – sprouting from a single base, or stool. Left alone it won't live much beyond a hundred years but if cut regularly – coppiced – it can survive almost indefinitely. The leaves are similar to those of the alder and the birch, round and serrated tapering to a tip; hazel actually belongs to the birch family, the Betulaceae.

Hazel's precious nuts, so vital to the life of the forest, grow from the fertilised female flowers – small red buds that appear on the branch alongside the male catkins. Rodents and birds feed on the buds while the catkins are an important food for insects. The catkins are produced the previous winter and sealed with resin until the spring. The flush of pollen following the melting of the resin is among the first foods of spring for insects waking from hibernation. The tree's smooth bark is also a vital host for many species of lichen.

'Haessel' comes from the Anglo-Saxon word for 'hood', the hat the nut wears. The Celts called hazel *Coll* and attributed wisdom to it. In Celtic mythology, nine sacred hazel trees grew around a pool, dropping nuts into the water to be eaten by a salmon. The spots on salmon corresponded to the number of nuts they had eaten. Hazel rods are used for divining water sources, while the smoke from burning nuts was part of an ancient and powerful Celtic ritual to divine the future.

In a world facing food shocks, the nuts of the hazel may once again become a critical part of the human diet.

Corylus avellana, common hazel

Juniperus, juniper

Juniper is the carpet of the boreal. Sometimes called mountain yew in Scotland, the tree creeps along the ground, hugging the earth, eschewing the idea of a tree in many respects. It is the most widely distributed evergreen conifer in the world, found from Japan to Europe and Africa and in North and Central America. There are many variants in the boreal and around sixty species worldwide, all members of the medicinal family called Cuppressaceae.

The pale purple berry of the female plant is a spice used for flavouring meat and gin but was also a medicinal ingredient. It emerges from within a guard of short spiky leaves like scales which are heat resistant and water repellent and contain toxins to discourage browsing. The waxy cuticles promote ecological health in three ways: they shade the ground, preserving soil moisture; they control soil erosion; they concentrate medicinal resins for birds and for soil and atmospheric health generally.

The birds of the boreal, especially the fieldfare and ring ouzel, that eat the juniper appear to be crucial to its germination. The berry hangs on the plant for up to three years. Inside its impermeable coat, the seed needs the striation that comes from passing through a bird's stomach in order to germinate. The plants and the birds need each other.

Juniper's resinous branches have always had a sacred role, and smoke from burning them is integral to many indigenous ceremonies. For Native American tribes juniper is a symbol of protection. Plains Indians such as the Cheyenne and Dakota hung boughs of juniper on their teepees to protect them from storms. The resins are antiviral and health giving, assisting with respiration and breathing problems. Gaelic folklore suggests juniper can also assist in inducing contractions of the uterus to bring on childbirth.

Although found all over the world, in certain places juniper is declining fast and is in need of special protection. The tree doesn't like shade, and the medicines in it need sunshine to be activated and released. It would be beneficial to plant juniper around schools, nurseries, care homes and hospitals.

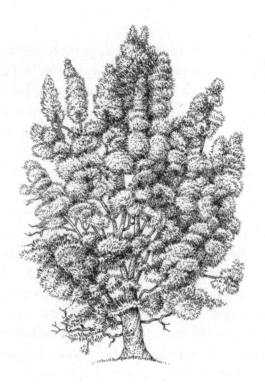

Juniperus communis, common juniper

Larix, larch

Known as tamarack in North America, the larch is an outlier, a deciduous conifer. There are only nine species of larch worldwide, with most found in Siberia and the hinterland of Eurasia. *Larix laricina* is the tamarack of Canada and Alaska, also called the weeping larch. Taller and more elegant than the stunted survivors of the Siberian cold, *Larix gmelinii* and *sibirica*, it graces woods from the Rockies to the Atlantic coast.

Larch and water go together, not only because of its ability to regulate groundwater flows and transform water from liquid to ice and back again through the gymnastics of its cells, but also because of how it reproduces. Larch pollen, like that of other conifers, travels through water. The sperm swims just like a human sperm up the pollen tube to fertilise the egg.

The deciduous mechanism of the larch makes it the most unusual conifer. The production of abscisic acid is what sets it apart. This hormone encourages the flow of chlorophyll out of the needles of the tree, causing them to turn orange or brown. A further drop in the autumn temperature causes another flush of abscisic acid which attacks the petiole tissue that attaches the needle to the branch. Leaf litter accumulates, and the tree is dormant for winter. In spring the trees switch on again, inhaling carbon dioxide, which they use to produce leaves which subsequently fall. This process captures megatons of carbon, which is stored because the shade from the trees cools the forest floor, slowing down fungal activity, decreasing evaporation and inhibiting decomposition.

For the indigenous peoples of Siberia, the larch was the tree of life, central to all their mythology and sacred rituals. By the peoples of North America and Europe, tamarack was prized for its roots, which were used to stitch birch-bark canoes together and even to join the decks and hulls of large sailing ships. Pipes made from straight tamarack trunks – which resist rot – were used in wells.

Larch is one of the most effective trees for sequestering carbon dioxide and regulating groundwater and so vital when devising strategic approaches to ecology.

Larix sibirica

Picea, spruce

Diana Beresford-Kroeger calls the spruce the workhorse of the global forest. There are forty-five species, but the pre-eminent two are the white spruce and black spruce, which 'launder' the bulk of the subarctic atmosphere.

Picea glauca is the white version, otherwise known as Canadian spruce, pasture spruce, cat spruce or, to the Chipewyan people, big brother. *Picea mariana* is the black, also called eastern, bog, swamp or double spruce.

Both have the firm green needles that are actually leaves rolled tightly into a tube, arranged spirally around the branch. Their trunks are long and straight, although northern specimens and black spruce with their toes permanently wet in the muskeg may be stunted. Needles, bark, roots and cones are all rich in gum and resins, which have great medicinal as well as pyrotechnic value. Slow-growing trees in the most marginal habitats are the richest in medicines.

The spruce's super-dark mesophyll can photosynthesise in extreme conditions and extracts the maximum value from tough corners of the planet. Its dark leaves absorb radiation as well as photosynthesising, cooling the planet twice by preventing long-wave radiation from returning out into the atmosphere only to be trapped by greenhouse gases.

In North American mythology spruce trees have varying significance. To south-western tribes, they are symbols of the sky. In one Hopi myth the spruce tree was a medicine man who transformed himself into a tree. In the Pima flood myth a father and mother of the Pima people survived the deluge by floating in a ball of spruce pitch. And there is an Iroquois legend about a girl rescued from witches by the spirit of the spruce.

By the Anishinaabe, spruce is called *ka-wa-tik*, and the cones of young trees were boiled to treat diarrhoea. Spruce roots, *wahtup*, were softened in water and used to tie birch bark together for canoes and as snares for rabbits.

Gum, pitch, resins and oils produced from spruce can be used instead of petroleum-based products. With a bit more research and attention, alternative industries are waiting to be discovered.

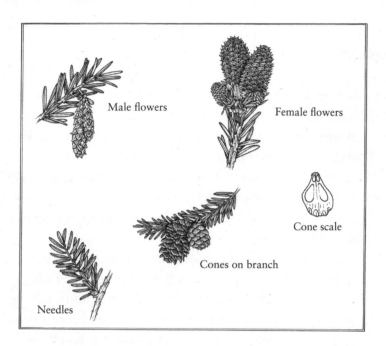

Male flowers   Female flowers

Cone scale

Cones on branch

Needles

Picea mariana, black spruce

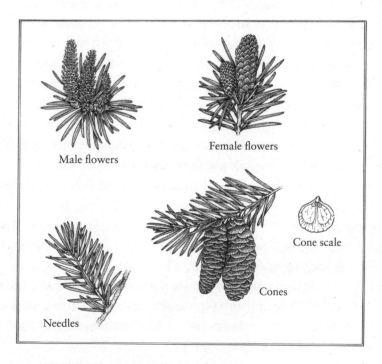

Female flowers

Male flowers

Cone scale

Cones

Needles

Picea glauca, white spruce

Pinus, pine

Pinus sylvestris, the Scots pine, is the pre-eminent representative of the pine family in Eurasia, and its twin, *Pinus banksiana*, holds sway over North America. Called variously jack pine, scrub pine, black pine and grey pine by the colonial settlers, its Native American name was *kohe*, an Athapsakan word. Another North American species, the bristlecone pine (*Pinus aristata*) is a candidate for the world's oldest tree, with specimens over 5000 years old in California.

The Scots pine lives longer and reaches higher than the jack pine, but the latter is tougher, surviving on bare rock and sandy soil, and it has developed an extraordinary relationship with lichen to access nutrients the thin soil cannot provide. These symbiotic relationships take place on the roots, but also the trunk, the twigs and needles. The picturesque lichen *Usnaceae*, old man's beard, can be found trailing off the branches of jack pines, capturing nitrogen and feeding it to the tree, and manufacturing a range of antibacterial acids and biochemicals.

The Scots pine produces pairs of blue-green needles and straight symmetrical cones which stand upright on the branch, but the jack pine produces stunted sharp needles and a pair of twisted female cones opposite each other on the branch, like coupled bananas. These cones are sealed with gum and only open in fire. The resin that means the firewood of the jack pine is prized for its high heat output also means that a dead pine is slow to decay, with snags persisting for up to a hundred years.

North America is rich in pines of all kinds, over one hundred species, and the indigenous peoples of the continent had a wide array of medicinal uses for them. The needles, resin and gum are antiseptic and antibacterial. Various parts of the tree were burned, steeped or boiled to aid in respiratory diseases. The Cayuga collected pine knots – where the medicine was most strongly concentrated – and extracted the pith to treat tuberculosis.

For the Great Lakes tribes, pines are a symbol of harmony with nature. The Iroquois burned white pine (*Pinus strobus*) needles to banish ghosts and seek peace, while fallen branches were burned and the smoke used to 'wash' the eyes of someone who had seen a dead person. In Siberia pine groves around the shores of Lake Baikal were sacred to the Buryat. In the UK Scots pines were traditionally used as markers in the landscape, denoting borders and rights of way. The ancient Egyptians buried an image of the god Osiris in the hollowed-out centre of a pine tree.

While the Scots pine and white pine are both vulnerable to drought, the jack pine seems to cope very well with seasonal water shortages. Given its huge range across North America extending over 1000 miles south of the Artic Circle, it could be a candidate for future forests with its wide climate niche.

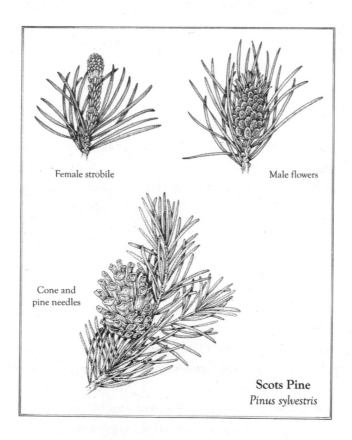

Female strobile

Male flowers

Cone and
pine needles

Scots Pine
Pinus sylvestris

Pinus sylvestris, Scots pine

Populus, aspen, poplar

Aspen is one of the most widely distributed trees in the world, from the Arctic Circle to North Africa and across the boreal to Japan. Like birch, it is a pioneer species and was among the first to colonise the northern hemisphere after the last ice age. It is fast growing and regenerates vigorously after disturbance or fire.

The bark is smooth and grey, its leaves small and round. They are uniquely adapted at the base of the petiole, where they flatten and are very flexible yet strong, being able to rotate and flutter in the breeze, giving rise to the name trembling or quaking aspen – *Populus tremuloides*. The Irish said it was shaking with emotion. When they first appear the leaves are coppery brown, then fill with chlorophyll and turn green, before going yellow in autumn. The quivering of the leaves reflects light around the tree and distributes the biochemicals in the leaf.

Aspen is an important source of medicine for animals and humans alike. Butterflies come to it for its salicylates and minerals such as zinc and magnesium. The low acidity of the bark means it is home to some lichens that only grow on aspen. Humans can eat the inner bark, which tastes akin to melon, and the outer bark has been used to treat everything from diabetes, heart disease and venereal disease to stomach ache. The leaves can relieve the pain of bee and wasp stings, and the white periderm of the bark can be collected and applied as a powder to stem bleeding.

Aspen flowers in spring produce long threads of downy seeds like cotton, hence the generic name cottonwood. The seeds are rarely used, the tree preferring to reproduce vegetatively, cloning itself. Large stands of aspen as in Poplar River are likely a single organism thousands of years old. The Anishinaabe used poplar for smoking food and hides as well as burning smudges to keep mosquitoes at bay.

In Greek myths it was known as the shield tree, offering physical and spiritual protection. The tree's Gaelic name is *critheann*, and the Highlanders thought it magical, associated with the faerie realm. As such, there was a taboo on using it for construction.

Globally, the aspen appears to be suffering from heat stress. However, on both sides of the Atlantic, Diana Beresford-Kroeger notes, *Populus tremuloides* has produced sports with triploids – three sets of chromosomes instead of the normal two – which could offer the chance to breed more resilient specimens.

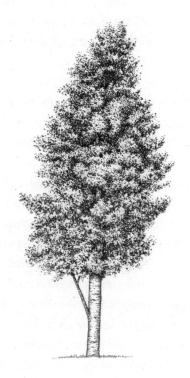

Populus tremuloides, trembling aspen

Populus balsamifera, balsam poplar

Balsam poplar is another cottonwood of the willow family, the Saliaceae, with clouds of fluff released by the female trees every spring. Like the aspen, the balsam poplar has a habit of regenerating vegetatively, sending out suckers underground up to forty metres from the mother tree. It looks similar to the aspen with a straight grey trunk, but although smooth in young saplings, this will crack and fissure into deep grooves with time. The leaves are longer, greener and somewhat heart shaped, much larger than those of its aspen cousin.

The balsam of the name comes from the rich oleoresin found in the tree's buds and leaves. This is the medicine cabinet of the tree, and the source of one of its nicknames, the balm of Gilead. This resin is concentrated because of the tree's preference for the harsh habitat of the treeline zone. Where very few hardwoods can survive, the balsam poplar thrives, enduring the cold and growing up to thirty metres tall and two metres in diameter. It is never found in krummholz, stunted form, but only ever grows straight and tall.

Although found all over the world, it does not form large stands outside North America. It was an important tree for the First Nations who called it the bam, bamtree or hackmatack, and who used the resin to treat a wide range of illnesses including cancer, high blood pressure and heart disease. The oil of the tree is a cure-all medicine used in many different treatments by indigenous Americans, and the wood has medicinal, antibacterial properties and so is used for toothpicks and chopsticks. Western medicine has yet to catch up and fully investigate the medicinal potential of the balsam poplar.

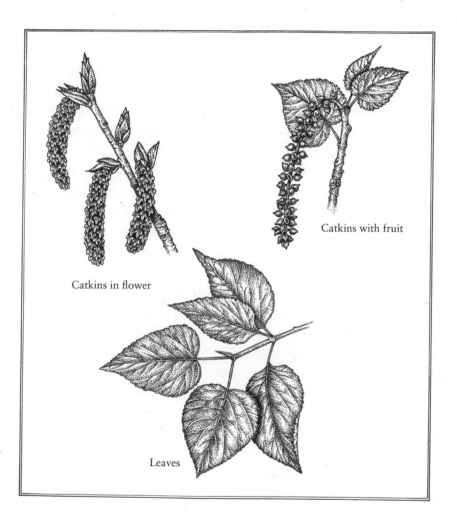

Catkins with fruit

Catkins in flower

Leaves

Populus balsamifera, balsam poplar

Salix, willow

Willows are the frontier species, found beyond the treeline, marking its transition to tundra. As eared willow or goat (pussy) willow it can grow as a shrub, creeping along the ground, or as crack or white willow a mature tree of up to thirty metres in height. There are 300 species stretching over a very wide climate niche. All of them share a love of water.

The willow hugs waterways, lining lowland rivers, tundra ponds and also mountain streams above the treeline. Being so close to water, there is a risk of mildew, but the willow has evolved a allelo-chemical called salicylanilide with anti-fungal and anti-mildew properties. Willow protects the upper reaches of watersheds, managing groundwater flow, slowing flooding and releasing restorative biochemicals into the water to the benefit of fish and other aquatic life. Labile esters help oil fixation in fish. Other salicyclic acids amplify light in water to the benefit of aquatic plants, and willow seed, once it has shed its cotton mane, floats in the river and turns a golden colour, tiny protein-packed food parcels for lucky fish.

In winter the bare branches of most species are colourful – green, yellow and red – and in spring catkins appear before the first leaves – chunky, furry, hairy buds, well insulated and attractive. The willow family feeds over 450 species of insects.

Its catkins are among the first sources of pollen for bumblebees. Bees will search for willows in order to access its medicinal pollen and nectars, which contain antibiotic properties. Butterflies too rely on willows for biochemicals to chelate the metals they need for their spectacular colours. Metals are the electron core in the colour of their wings.

Willows are critical to clean and healthy watersheds but also to a healthy insect population. As pollinators struggle with global warming, planting more willows will help them. Most species will grow from a cutting pushed into the ground. Willow seeds start growing thirty-six hours after germination and don't stop: they are among the fastest-growing trees of the forest.

The young shoots have always been used for baskets and weaving, and the mature wood is good for many things including furniture, cartwheels, cricket bats and clogs. The Ojibwe make their sweat lodges out of willow, and its medicinal uses range from pain relief and combating inflammation to relieving constipation and as an antibiotic. The twigs of willow have long been chewed since before the active ingredient aspirin was isolated and marketed to the world.

Salix fragilis, crack willow

Sorbus, rowan

All across the boreal, the rowan is a familiar sight in spring with its bloom of white flowers and in autumn with its crop of bunched red berries. A member of the rose family, the Rosaceae, its symmetrical serrated leaves resemble those of the ash, giving rise to its other name, mountain ash. *Sorbus aucuparia* is the common European form, *S. americana* and *S. decora* North American variants. The Greenland mountain ash, *S. groenlandica*, is its own subspecies.

The *Sorbus* genus are fast-growing pioneers of the forest, easy to establish in unlikely places and therefore a key member of the boreal club, with useful relations with fungi and lichen second only to hazel. The spring flowers have a strong sweet scent attractive to pollinators and are a feeding ground for all manner of insects, providing food for migratory birds showing up after a long winter. The rowan is the safety net of the boreal, managing to produce pollen and nectar regardless of disruptions in weather patterns.

Then in autumn the birds feast again – on the red berries of the *Sorbus* – before flying south. The red of the rowan is the harbinger of winter. The birds in turn disperse the seeds; eight are contained within each red globe of fruit. The tough coat of the seed requires work to break it down, either in the digestive tract of an animal or through the actions of the weather. Seeds can germinate several years after they have been produced.

Humans too can eat rowan berries, and preserves were often made to be eaten with meat. In the Highlands of Scotland there was a taboo on using any part of the tree save the berries, and a strict taboo on cutting the wood with a knife. Rowans were planted near homes for protection, and were associated with the faerie kingdom. The tiny five-pointed star of the berry is a pentagram, an ancient protective symbol. In Scandinavia runes were written on rowan wood for divination.

The leaves of rowan are silvery on the underside. They reflect light and can maintain and capture rising humidity from the soil. The root mass is spreading and can survive moisture deficits in winter and summer, and so rowan is a good tree to plant to mitigate climate change.

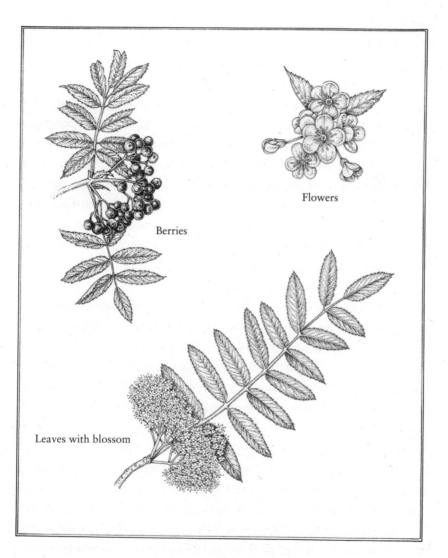

Berries

Flowers

Leaves with blossom

Sorbus aucuparia, rowan

Taxus, yew

The yew is a curiosity. A conifer fond of marginal soils in wet climates, it is present from the UK across Europe to mountainous areas of North Africa, Iran and the Caucasus, yet it is only found in small populations or as single trees, hardly ever as a complete forest. It is listed as endangered in many countries.

One European yew (*Taxus baccata*) is the oldest living tree in Europe – the Fortingall yew at Glen Lyon in Scotland – and *Taxus* is the oldest European tree genus, apparently emerging during the transition between the Cretaceous and the Tertiary, sixty-six million years ago. It is therefore difficult to look at the yew without respect; it commands attention because of its age and potential for immortality. This was the reason the Celts worshipped it as the tree of life and death. The red berries of the female tree are highly poisonous to humans although sometimes used to treat headaches and neuralgia. Lately, anti-cancer properties have been discovered. The trunk can be enormous, and the tree's low-sweeping branches, creating thick dark shade allowing very little to grow beneath them, give the yew an air of mystery and magic.

The yew is a reminder of how little we know about trees and how they got where they are. The yew seems to be a relict species – it was once much more widespread – but climate changes in the past have shaped its present distribution, restricting it to refugia. Evidence suggests that ice oscillations in the Neogene (thirty-four to thirty-five million years ago) degraded the distribution of the yew, and the oscillations of the multiple Quaternary glaciations, including the latest ice age, further fragmented the populations.[1]

The yew's dispersal ability is low. The tiny green ovules that, when fertilised with pollen, become the red berries are located between the leaf stalk and the stems of the leaves. The flat, needle-like leaves have grey and yellow bands underneath them and are attached to the branch in spirals, although a twist at the base gives them the appearance of running in rows. The yew does not disperse its seed far and does not grow from seed easily, needing a moist, nutrient-rich microclimate and a nurse plant, often juniper, to protect the seedlings from herbivores. Yew will outcompete hardwoods to create new forest on disturbed sites but, once felled within a mixed forest, will struggle to regenerate.

The yew remains from a phase of the forest before humans were on the scene. Its distribution now is a result of a climate that has turned against it and lost the humid mists of the Pliocene, as well as later felling by humans. In that respect, it is not just a ghost of the past, but the ghost of the future forest too.

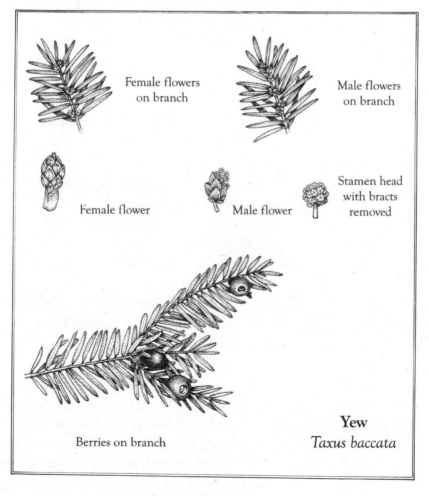

Female flowers on branch

Male flowers on branch

Female flower

Male flower

Stamen head with bracts removed

Berries on branch

Yew
Taxus baccata

Taxus baccata, yew

Notes

Prologue

1 Thomas Berry, *The Dream of the Earth* (Sierra Club, 1988)

1. The Zombie Forest

1 Ron Summers, *Abernethy Forest: The History and Ecology of a Scottish Pinewood* (RSPB, 2018)
2 Ibid.
3 Oliver Rackham, *Trees and Woodland in the British Landscape* (Phoenix, 1976)
4 Rob Wilson et al., 'Reconstructing Holocene Climate from Tree Rings: The potential for a long chronology from the Scottish Highlands', *The Holocene* 22, 3–11, 2019. See also Miloš Rydval et al., 'Spatial reconstruction of Scottish summer temperatures from tree rings', *International Journal of Climatology* 37:3, 2017
5 Jurata Buchovska and Darius Danusevicius, 'Post glacial migration of Scots pine', *Baltic Forestry*, 2019
6 Garrett Hardin, 'The Tragedy of the Commons', *Science* 162:3859, 1243–48, 13 December 1968. Hardin's argument was that humans cannot be trusted to act with temperance in the commons. The full phrase he used was 'the tragedy of freedom in the commons', and the article discusses ways in which humans might be compelled to exercise restraint and not over-exploit or pollute common areas. While this is a relevant argument to the management of natural resources now, it is not a useful explanation of the past nor a rigorous engagement with indigenous practices. However, that has not stopped Hardin's article being used in that way. See George Monbiot, 'The Tragedy of Enclosure', *Scientific American*, January 1994.
7 John Prebble, *The Highland Clearances* (Penguin, 1969)

8 Arthur Mitchell (ed.), 'Geographical Collections', 2 in Professor T. C. Smout, *History of the Native Woodlands of Scotland 1500–1920* (Edinburgh University Press, 2008)

9 Jim Crumley, *The Great Wood: The Ancient Forest of Caledon* (Birlinn, 2011)

10 Vladimir Gavrikov and Pavel Grabarnik et al., 'Trunk-Top Relations in a Siberian Pine Forest', *Biometrical Journal* 35, 1993

11 Diana Beresford-Kroeger, *The Global Forest: 40 Ways Trees Can Save Us* (Particular Books, 2011)

12 Rackham, *Trees and Woodland*

13 Eurostat database, ec.europa.eu

14 Leif Kullman, 'A Recent and Distinct Pine (Pinus sylvestris L.) Reproduction Upsurge at the Treeline in the Swedish Scandes', *International Journal of Research in Geography* 4, 2018

15 Leif Kullman, 'Recent Treeline Shift in the Kebnekaise Mountains, Northern Sweden', *International Journal of Current Research* 10:01, 2018

16 Summers, *Abernethy Forest*

17 Ibid.

18 Fiona Harvey, 'London to have climate similar to Barcelona by 2050', *Guardian,* 10 July 2019

19 Summers, *Abernethy Forest*

20 Bob Berwyn, 'Many Overheated Forests May Soon Release More Carbon Than They Absorb', *Inside Climate News*, 13 January 2019

2. Chasing Reindeer

1 *Last Yoik in Sami Forests?* Greenpeace, 2005. A film which documents the struggle against the clear-cutting of Finland's old-growth forests.

2 Personal communication from Diana Beresford-Kroeger

3 Diana Beresford-Kroeger, *Arboretum Borealis* (Michigan University Press, 2010)

4 Abrahm Lustgarten, 'How Russia Wins the Climate Crisis', *New York Times*, 9 December 2020

3. The Sleeping Bear

1 Anton Chekhov, *Sakhalin Island* (Alma Classics, 2019)

2 Anatoly Abaimov et al., 'Variability and ecology of Siberian larch species', Swedish University of Agricultural Sciences, Department of Siviculture, Report 43, 1998

3 Bob Berwyn, 'When Autumn Leaves Begin to Fall – As the Climate Warms, Leaves on Some Trees are Dying Earlier', *Inside Climate News*, 26 November 2020

4 Berwyn, 'Many Overheated Forests . . .'

5 Elena Parfenova, Nadezhda Tchebakova and Amber Soja, 'Assessing landscape potential for human sustainability and "attractiveness" across Asian Russia in a warmer 21st century', *Environmental Research Letters* 14:6, 2019

6 Lustgarten, 'How Russia Wins . . .'

7 Ibid.

8 Oliver Milman, 'Global heating pushes tropical regions towards limits of human livability', *Guardian*, 8 March 2021

9 Gabriel Popkin, 'Some tropical forests show surprising resilience as temperatures rise', *National Geographic*, 19 November 2020

10 A. A. Popov, *The Nganasan: The Material Culture of the Tavgi Samoyeds*, Routledge Uralic and Altaic Series 56, Routledge, 1966

11 Piers Vitebsky, *The Reindeer People: Living with Animals and Spirits in Siberia* (Mariner Books, 2005)

12 Peter Wadhams, *A Farewell to Ice* (Penguin, 2015)

13 Svetlana Skarbo, 'Weather swings in Siberia as extreme heat is followed by June snow, tornadoes and floods', *Siberian Times*, 9 June 2020

14 *Shaman*, Lennart Mari. A documentary film made in 1977 and released in 1997, https://www.youtube.com/watch?v=2ZlOPkIbR50

15 Eugene Helimski, 'Nganasan Shamanistic Tradition: Observations and Hypotheses,' Paper presented to the Conference 'Shamanhood: The Endangered Languages of Ritual' at the Centre for Advanced Study, Oslo June 1999.

16 W. Gareth Rees et al., 'Is subarctic forest advance able to keep pace with climate change?' *Global Change Biology* 26:4, April 2020

17 Dr Zac Labe of the University of Colorado interviewed by Jeff Berardelli, 'Temperatures in the Arctic are astonishingly warmer than they should be', *CBS News*, 23 November 2020

18 Chekhov, *Sakhalin Island*

19 Craig Welch, 'Exclusive: Some Arctic Ground No Longer Freezing – Even in Winter', *National Geographic*, 20 August 2018

20 S. Zimov et al., 'Permafrost and the global carbon budget', *Science* 312:5780, 16 July 2006

21 University of Copenhagen, 'Arctic Permafrost Releases More Carbon Dioxide than Previously Believed', phys.org, 9 February 2021

4. The Frontier

1 Charles Wohlforth, *The Whale and the Supercomputer* (Farrar, Straus & Giroux, 2004) details the story of this research and the debates surrounding it.

2 Ken Tape, 'Tundra be dammed: Beaver colonization of the Arctic', *Global Change Biology* 24:10, October 2018; Ben M. Jones et al., 'Increase in beaver dams controls surface water and thermokarst dynamics in an Arctic tundra region, Baldwin Peninsula, northwestern Alaska', *Environmental Research Letters* 15, 2020

3 Seth Kantner, *Shopping for Porcupine* (Milkweed Editions, 2008)

4 Anna Terskaia, Roman Dial and Patrick Sullivan, 'Pathways of tundra encroachment by trees and tall shrubs in the western Brooks Range of Alaska', *Ecography* 43, 2020

5 Merlin Sheldrake, *Entangled Life (*Bodley Head, 2020)

6 S. W. Simard et al., 'Net transfer of carbon between ectomycorrhizal tree species in the field', *Nature* 388, 1997; Ferris Jaber, 'The Social Life of Forests', *New York Times Magazine*, December 2020

7 'Satellites reveal a browning forest', NASA Earth Observatory, 18 April 2006

8 'Land Ecosystems Are Becoming Less Efficient at Absorbing CO_2', NASA Earth Observatory, 18 December 2020

9 Kate Willett, 'Investigating climate change's "humidity paradox"', *Carbon Brief*, 1 December 2020

10 Max Martin, 'Add atmospheric drying – and potential lower crop yields – to climate change toll', *Toronto Star*, 12 March 2021

11 T. J. Brodribb et al., 'Hanging by a thread? Forests and Drought', *Science* 368:6488, 17 April 2020

12 Jim Robbins, 'The Rapid and Startling Decline of World's Vast Boreal Forests', *Yale Environment* 360, 12 October 2015

13 Ibid.

14 Fred Pearce, *A Trillion Trees* (Granta, 2021)

15 Ibid.

16 David Ellison et al., 'Trees, Forests and Water: Cool Insights for a Hot World', *Global Environmental Change* 43, 2017

17 A. M. Makarieva and V. G. Gorshkov, 'Biotic pump of atmospheric moisture as driver of the hydrological cycle on land', *Hydrological Earth System Science* 11, 2007

18 Ibid.

19 Roger Pielke and Piers Vidale, 'The Boreal Forest and the Polar Front', *Journal of Geophysical Research* 100:D12, 1995

20 Makarieva and Gorshkov, 'Biotic pump of atmospheric moisture . . .'

21 Fred Pearce, 'A Controversial Russian Theory Claims Forests Don't Just Make Rain – They Make Wind', *Science*, 18 June 2020

22 Kyle Redilla, Sarah T. Pearl et al., 'Wind Climatology for Alaska: Historical and Future', *Atmospheric and Climate Sciences* 9:4, October 2019

23 Richard K. Nelson, *Make Prayers to the Raven: A Koyukon View of the Northern Forest* (University of Chicago Press, 1983)

24 'Project Jukebox', University of Alaska Fairbanks Oral History Program. See https://jukebox.uaf.edu/site7/interviews/3623 for the interviews with Attla.

25 *Make Prayers to the Raven*, KUAC Radio, Fairbanks. Documentary series available on YouTube.

26 World Wildlife Fund for Nature and Huslia Tribal Council, *Witnessing Climate Change in Alaska*, 2005. A student-led series of radio programmes interviewing residents of Huslia available at https://wwf.panda.org/discover/knowledge_hub/where_we_work/arctic/what_we_do/climate/climatewitness2/huslia/radio_programmes/

27 Juliet Eilperin, 'As Alaska warms, one village's fight over oil and development', *Washington Post*, 14 December 2019

28 Beresford-Kroeger, *Arboretum Borealis*

29 Ibid.

30 Dieter Kotte et al. (eds), *International Handbook of Forest Therapy* (Cambridge Scholars, 2019)

31 Sabrina Shankman, 'What Has Trump Done to Alaska? Not as Much as He Wanted', *Inside Climate News*, 30 August 2020

5. The Forest in the Sea

1 Diana Beresford-Kroeger, *To Speak for the Trees: My Life's Journey from Ancient Celtic Wisdom to a Healing Vision of the Forest* (Penguin, 2019)

2 Ibid.

3 John Laird Farrar, *Trees in Canada* (Fitzhenry and Whiteside, 2017)

4 Beresford-Kroeger, *To Speak for the Trees*; see also Katsuhiko Matsunaga et al., 'The role of terrestrial humic substances on the shift of kelp community to crustose coralline algae community of the

southern Hokkaido Island in the Japan Sea', *Journal of Experimental Marine Biology and Ecology* 241, 1999

5 Charles C. Mann, *1493: How Europe's Discovery of the Americas Revolutionized Trade, Ecology and Life on Earth* (Granta, 2011)

6 Tracy Glynn, 'Canada is under-reporting deforestation, carbon debt from clearcutting: Wildlands League', *NB Media Coop*, 15 January 2020; Frederick Beaudry, 'An Update on Deforestation in Canada', treehugger.com, 31 January 2019

7 The fire cycles and the cultural geography of the Anishinaabe is detailed in Iain J. Davidson-Hunt, Nathan Deutsch and Andrew M. Miller, *Pimachiowin Aki Cultural Landscape Atlas: Land That Gives Life* (Pimachiowin Aki Corporation, 2012)

8 David Lindenmayer and Chloe Sato, 'Hidden collapse is driven by fire and logging in a socioecological forest ecosystem', *Proceedings of the National Academy of Sciences of the USA* 115:20, 2018

9 Robin Wall Kimmerer, *Braiding Sweetgrass* (Milkweed, 2013)

10 *Pimachiowin Aki Cultural Landscape Atlas* explains the geology of Lake Agassiz in detail.

11 Columbia University, 'Northern peatlands may contain twice as much carbon as previously thought', phys.org, 21 October 2019

6. Last Tango with Ice

1 Wadhams, *A Farewell to Ice*

2 C. Rahbek et al., 'Humboldt's enigma: What causes global patterns of mountain biodiversity?' *Science* 365:6458, September 2019

3 Richard T. Corlett and David A. Westcott, 'Will plant movements keep up with climate change?' *Trends in Ecology and Evolution* 28:8, 2013

4 Jared Diamond, *Collapse: How Societies Choose to Survive or Fail* (Penguin, 2005)

5 Andrew Christ and Paul Bierman, 'Ancient leaves preserved under a mile of Greenland's ice – and lost in a freezer for years – hold lessons about climate change', *Conversation*, 15 March 2021

6 Ker Than, 'Ancient Greenland Was Actually Green', *Livescience*, 5 July 2007

7 Signe Normand et al., 'A Greener Greenland?: Climatic potential and long-term constraints on future expansions of trees and shrubs', *Philosophical Transactions of the Royal Society* 368:1624, 2013

8 Peter Branner, 'The Terrifying Warning Lurking in the Earth's Ancient Rock Record – Our climate models could be missing something big', *Atlantic*, March 2021

9 Max Adams, *The Wisdom of Trees* (Head of Zeus, 2014)

10 Ellison et al., 'Trees, Forests and Water . . .'

11 Aslak Grinsted and Jens Hesselbjerg Christensen, 'The transient sensitivity of sea level rise', *Ocean Science* 17, 2021

Epilogue

1 Ian Rappel, 'Habitable Earth: Biodiversity, Society and Re-wilding', *International Socialism*, April 2021

2 Ibid.

3 Donna Harraway, *Staying with the Trouble: Making Kin in the Chuthulucene* (Duke University Press, 2016)

4 James Hansen, *Storms of my Grandchildren* (Bloomsbury, 2009); James Hansen et al., 'Young people's burden: Requirement of negative CO2 emissions', *Earth System Dynamics* 8, 2017; David Wadsell, 'Climate Dynamics: Facing the Harsh Realities of Now, Climate Sensitivity, Target Temperature and the Carbon Budget: Guidelines for Strategic Action', presentation by the Apollo-Gaia Project, September 2015

5 David Abram, *The Spell of the Sensuous* (Vintage, 1997)

6 'Collaborative survival' is a phrase coined by Anna Tsing in *The Mushroom at the End of the World* (Princeton University Press, 2015)

Glossary

1 P. A. Thomas and A. Polwart, '*Taxus baccata*', *Journal of Ecology* 91, 2003

Acknowledgements

Thank you to my generous hosts and friends in so many places who helped me see with new eyes: in Scotland, Thomas MacDonnell, Margaret Bennett, Rob Wilson, Fiona Holmes and Trees for Life, Mark Hancock and Cairngorms Connect; in Finnmark, Hallgeir Strifeldt, Tor Havard Sund, Māret Buljo, Inge-Marie Gaup Eira, Issát H. Eira, Berit Utsi, Niillas Mihkkael, Mārija Eira, Sara-Irene Haetta, Thomas Myrnes Nygård; in Russia, Elena Kukovskaya, Nadezhda Tchebakova, Aleksandr Bondarev, Sophy Roberts, Ko van Huissteden, Sander Veraverbeke, Djazda and Maria Yevstappi, Misha and Anna Chuperin, Anatoly Gavrilov, Nikolai Zimov, Nikolai Kozak and Nikolai Baronofsky; in Alaska, Pat Lambert, Adam Weymouth, Ken Tape, Seth Kantner, Roman Dial, Patrick Sullivan, Rebecca Hewitt, Brendan Rogers, John Gaedeke, Carl Burgett; in Canada, Diana and Christian Beresford-Kroeger, Sophia and Ray Rabliauskas and everyone in Poplar River, LeeAnn Fishback, Steven Mamet, Dave Daley; in Greenland, Kenneth Høegh, Ellen Friedrikssen, Dirk van As, Faezeh M. Nick, Peter Friis Møller, Jason Box and all at Greenland Trees, and to Jennifer Cartwright for an education in refugia.

Thank you to the wonderful team in London and New York who championed the project from the outset: Sophie Lambert, Anna Stein, Bea Hemming and Elisabeth Dyssegaard. To Jonny Donovan, whose companionship and enthusiasm got me started in Norway. To early readers and critical friends Jay Griffiths, Tom Bullough, Diana Beresford-Kroeger, Elfie Rawlence, Simon Rawlence and Zanna Jeffries. Thank you to the trees that gave their lives for this book. To my wife Louise and daughters Daphne and Cissy, who have seen less of me than they should. And lastly, to Brigid Hogan for commissioning the beautiful illustrations by Lizzie Harper. The drawings are dedicated to the memory of her brother and my father-in-law, Patrick Hogan, an oak of a man who passed away during the pandemic while this book was being written.

Index

Page references in *italics* indicate images.